Grassland use in Europe

A syllabus for young farmers

Agnes van den Pol-van Dasselaar, Leanne Bastiaansen-Aantjes,
Fergus Bogue, Michael O'Donovan, Christian Huyghe, eds

Éditions Quæ

This project has received funding from the European Union's Horizon 2020 research and innovation programme under grant agreement No 727368.

Shared Innovation Space for Sustainable Productivity of Grasslands in Europe

Project Acronym: Inno4Grass

Deliverable 5.3. Specific grassland syllabus (D5.3.1) and PowerPoint presentations (D5.3.2) available for young farmers and advisors

Responsible partner: Teagasc (WP Leader) and Aeres (Task Leader)

Other contributors: GLZ, WUR, RHEA, IDELE, APCA, LWK, UGOE, TRAME, AWE, SLU, NLTO, CNR, PULS, WIR, AIA, LRC

Submission date: 28 February 2019

ISBN: 978-2-7592-3145-4
e-ISBN (pdf): 978-2-7592-3146-1
x-ISBN (ePub): 978-2-7592-3147-8

Preface

Agnes van den Pol-van Dasselaar, Leanne Bastiaansen-Aantjes, Fergus Bogue, Michael O'Donovan, Christian Huyghe

The future of grassland farming in Europe is in the hands of young farmers. Compared to other topics, grassland management has often been a weak point of teaching delivered by agricultural technical schools in several European countries. The training of future farmers and advisors could thus be significantly improved, which could lead to better grassland management in the medium and long terms. For this reason, the European project Inno4Grass created a syllabus and a set of PowerPoint presentations on practical grassland management for current and future generations of grassland farmers and advisors. The PowerPoint presentations can be downloaded from Encylopedia pratensis (www.encyclopediapratensis.eu) in the different languages of the countries participating in Inno4Grass. The syllabus and the PowerPoints contain the necessary practical and technical knowledge required for sustainable grassland management.

The syllabus is written by a group of authors. By combining the expertise of experts from the different Inno4Grass partner countries, we were able to create a document that provides both general knowledge and country-specific information and examples. Every part of this syllabus was written by at least two authored and reviewed by at least two other individuals, usually from different countries, to ensure that all relevant and available information was included. This process also ensured that the text provides examples from various European countries. The authors of specific parts of the syllabus are mentioned at the beginning of the relevant sections. A full list of authors can be found on the following page. The preface of this syllabus contains additional general information about Inno4Grass as well as the state of the art of European grasslands. Thereafter, the important general aspects of grassland management are presented in the different chapters: grassland production (Chapter 1), grazing management (Chapter 2), hay and silage making (Chapter 3), soil and nutrient management (Chapter 4), environment and biodiversity (Chapter 5) and quality of products from grass (Chapter 6). The syllabus ends with specific information on characteristics of the individual countries participating in Inno4Grass (Chapter 7).

We hope and expect that this syllabus will be a source of inspiration for all (future) farmers and advisors.

List of authors

Sweden	Nilla Nilsdotter-Linde	Swedish University of Agricultural Sciences
	Eva Spörndly	Swedish University of Agricultural Sciences
	Rolf Spörndly	Swedish University of Agricultural Sciences
Ireland	Fergus Bogue	Teagasc
	Michael O'Donovan	Teagasc
The Netherlands	Agnes van den Pol-van Dasselaar	Aeres University of Applied Sciences / Wageningen University
	Hein de Kort	LTO
	Leanne Bastiaansen-Aantjes	Aeres University of Applied Sciences
Belgium	Alain Peeters	RHEA Research Centre
	Benoît Delaite	TR@ME
	Xavier Delmon	TR@ME
	Daniel Jacquet	AWE: Association Wallonne de l'Élevage
	Patrik Gauder	AWE: Association Wallonne de l'Élevage
Germany	Johannes Isselstein	University of Göttingen
	Martin Komainda	University of Göttingen
	Nora Schiebenhöfer	University of Göttingen
	Felicitas Kaemena	LWK: Landwirtschaftskammer Niedersachsen
	Talea Becker	Grünlandzentrum
	Lena Dangers	Grünlandzentrum
	Arno Krause	Grünlandzentrum
Poland	Piotr Goliński	Poznan University of Life Sciences
	Barbara Golińska	Poznan University of Life Sciences
France	Fanny Sauvaire-Baste	APCA: Assemblée Permanente des Chambres d'Agriculture
	Julien Fradin	IDELE: Institut de l'Élevage
	Christian Huyghe	INRA
Italy	Giovanni Peratoner	Laimburg Research Centre
	Riccardo Negrini	Associazione Italiana Allevatori
	Lorenzo Pascarella	Associazione Italiana Allevatori
	Rita Melis	ISPAAM-CNR
	Claudio Porqueddu	ISPAAM-CNR

Table of contents

CHARACTERISTICS
OF INDIVIDUAL COUNTRIES

Objectives of Inno4Grass

Arno Krause and Talea Becker

European farmers, and especially grassland farmers, are facing tremendous challenges. Not only must they deal with instable prices for milk and meat coupled with rising input prices, but they must also take into consideration high societal demands with regard to environmental protection and animal welfare standards. Grasslands are vitally important both for agriculture and society. Permanent and temporary grasslands cover 61 million ha across the EU-28, which accounts for 16% of the total EU land area and 40% of the EU agricultural area (Eurostat, 2010). These grasslands serve multiple functions, including the local provision of fodder for animal husbandry (and hence high-quality food provision for people), biodiversity conservation, carbon storage and the provision of 'traditional' landscapes that European citizens appreciate for recreational purposes and cultural heritage. The large diversity of management practices, soils and climates enhance the range of ecosystem services provided by grasslands. Farmers in the EU often do not perceive the multi-functionality of grasslands as an advantage. This leads to undervaluation and a lack of valorisation strategies. Since market-oriented concepts to create rewards for ecosystem services have not yet been sufficiently understood or developed, their multi-functionality increasingly turns grasslands, especially in intensive production systems, into areas of conflict between food demand and calls for the provision of other ecosystem services. The potential for better use of grasslands to reduce production costs in livestock farming has also been underestimated.

To cope with the challenges and better capitalise on the advantages of grasslands, farmers need dedicated innovations that help them improve grassland economic performance. It is important for these innovations to not only have been developed and tested in scientific institutions, but that they have also been used on practical farms and are adapted to the conditions on farms.

Several reasons explain the low adoption of innovations in grasslands:
– Grassland-based production systems are complex and therefore innovative systems must be implemented as a combination of innovative practices that take local conditions into account.
– Innovation benefits from grasslands are not immediately apparent and require patience.
– Grassland innovation affects various aspects of sustainability (profitability, environment, social acceptance), often in a contradictory manner.
– There is limited to no interaction between practice and research.

Despite these issues, some farmers are very innovative – especially early adopters, who develop solutions and techniques on their own or adapt existing ideas to the conditions of their farm. It takes time for these innovations and techniques to spread among the community of farmers, and sometimes they do not reach the entire community. Collaboration between farmers, advisors and scientists is insufficient in the countries concerned. For this reason, the latest research results are not sufficiently applied and valuable knowledge related to grasslands is discovered by practitioners at a late stage.

The aim of Inno4Grass is to overcome these issues and to foster the gathering, spreading and creation of knowledge. The overall objective of Inno4Grass is to close the gap between practice and science, to identify innovations that would otherwise be ignored and to ensure the implementation of innovative systems on productive grasslands. The project's long-term goal is to increase the profitability of European grassland farms and to preserve environmental values. Apart from identifying innovations, Inno4Grass facilitates the process of knowledge co-creation. For this, Inno4Grass brings together farmers, scientists, advisors and teachers to develop solutions for typical problems and to adapt existing ideas to practical farms. These groups are moderated by facilitator agents, which facilitate communication and exchange between members of the groups. They use a participatory approach to initiate discussions on innovation using online discussion groups and farmer networks. Inno4Grass is an international project that fosters the exchange of ideas within eight European Member States: Belgium, France, Germany, Ireland, Italy, Poland, Sweden and the Netherlands. Within the project there are meetings of facilitator agents and exchange visits of farmers between countries to support the cross-border flow of information, thereby helping the project to benefit from diversity amongst farmers.

The innovations identified through the project are shared with the community of farmers. They are documented with farm portraits and can be found on the project homepage (https://www.inno4grass.eu/en/dissemination). Along with this syllabus, additional dissemination materials are available (innovations, abstracts, video clips, leaflets) and used for the enrichment of national and European Wikimedia and the Encyclopedia pratensis (https://www.encyclopediapratensis.eu).

Kick-off meeting for the Inno4Grass project, Berlin.

Material for training available at:
https://www.encyclopediapratensis.eu/encyclopedia_pratensis/educational-resources/
material-for-training/

Diversity of European grasslands

Alain Peeters and Johannes Isselstein

The vast majority of European grasslands are man-made. They have developed concomitantly with livestock husbandry. European grasslands are extremely diverse with regard to their management, soil types, plant composition, production potential and fodder quality. They can be divided in two main categories: permanent and temporary grasslands (Peeters *et al.*, 2014).

▶▶ Main grassland types

Permanent grasslands are grasslands that have not been completely renewed after destruction for ten years or more. They can be agriculturally-improved, semi-natural, natural or no longer used for production. They can be species-rich or species-poor. They comprise grasses, legumes, forbs and grass-like plants in variable proportions. Trees and shrubs can be present, such as in grazed wooded areas.

Agriculturally-improved permanent grasslands are located on soils with a moderate or high fertility that allow for intensive agricultural management. Compared to semi-natural grasslands, fertilisation and stocking rates are higher, and the swards are defoliated more frequently and have a higher herbage and livestock production.

Natural and semi-natural grasslands are low-yielding permanent grasslands, dominated by indigenous, naturally occurring grass communities, other herbaceous species and, in some cases, shrubs and/or trees. These mown and/or grazed ecosystems have not been substantially modified by agricultural practices. Natural grasslands are rather rare in Europe and occur spontaneously in marginal environments such as mountain tops, tundra or salty soils. Semi-natural grasslands are associated with human activities; indeed, without human intervention most of them would be colonised by woody vegetation.

Temporary grasslands are grasslands that are sown with annual, biennial or perennial forage species. They are sown on arable land and can be integrated in crop rotations or sown after a preceding grass crop. They are kept for a short period of time, from a couple of months to usually a few years. They are generally established with pure sowings of legumes, pure sowings of grasses or grass/legume mixtures.

▸▸ Economic and social importance of European grasslands

Permanent grasslands are an important component of European landscapes and farming systems. They cover about 60 million ha in the European Union (EU-28, 2013), while temporary grasslands cover about 11 million ha. Together, they occupy about 40% of the European utilised agricultural area (UAA) (Eurostat).

These grasslands are the feeding basis of about 196 million head of grazing livestock. They are managed by about 3.6 million holders, i.e. about 33% of all European farm managers (Eurostat: EU-28 in 2013). There were about 134 million livestock units (LSU) of total livestock and 78 million LSU of grazing livestock (59%) in the EU-27 in 2007, with the vast majority of livestock located in the EU-15. For the grazing livestock population (in LSU) in the EU-27, 82% were cattle and 14% small ruminants (sheep and goats). Dairy cows accounted for 31% and other cows (mainly suckler cows) for 16% of the total LSU of grazing livestock; two-thirds of cows were thus dairy cows and one-third other cows. Beef and veal, sheep and goat meat totalled 11% and milk 14% of total agricultural production value (Eurostat).

Grasslands are essential for feeding livestock, which then supply milk and meat to human populations. They are the cheapest source of feed to supply grazing livestock and can thus contribute to reducing production costs. Grass-fed milk and meat have unique nutritional properties for consumers that are sometimes highlighted by certified trademarks, such as 'Pasture for Life' in the United Kingdom.

Milk can be processed on farms into a wide range of products such as butter, cheese, yoghurt and ice cream. Meat can also be processed, usually by butchers, and sold by farmers as meat parcels or delicacies. These products can be sold through short and local marketing chains, which has the potential to significantly increase farmers' income and create jobs in agriculture.

▸▸ Environmental importance

Over the centuries, agriculture in Europe has created a patchwork of habitats that are very favourable to biodiversity. For instance, extensive grazing on common lands and haymaking in meadows have created diverse semi-natural ecosystems and attractive landscapes. These semi-natural grasslands are among the most species-rich habitats on the continent. Even intensively used permanent grasslands, although less diverse in terms of plants and insects, provide more ecosystem services than arable crops. Grasslands play a very important role in conserving European biodiversity, creating attractive landscapes (including for tourists), storing carbon in soils, improving soil fertility, and protecting soils from erosion and surface and groundwater from nitrate and pesticide pollution. They are also increasingly important in addressing climate change challenges as they offer the possibility of i) mitigating climate change effects thanks to a stable carbon storage and ii) limiting greenhouse gas emissions, and especially nitrous oxide (N_2O), as legumes may be used as the main source of nitrogen for plant growth, including in grass-legume mixtures.

▸▸ Threats

Permanent grasslands are threatened. In the EU-6 (Benelux, France, Germany, Italy), permanent grassland losses have been estimated at about 30% and 7 million ha between 1967 and 2007 (Eurostat). In some places, the losses have been even more substantial; for example, in Upper Normandy (France), the permanent grassland area fell by about 50% between 1970 and 2000.

Chapter 1

Grassland production

*Alain Peeters, Nilla Nilsdotter-Linde, Giovanni Peratoner,
Martin Komainda and Johannes Isselstein*

▸▸ Characteristics of grass

A. Peeters and N. Nilsdotter-Linde

Grass species are a major component of most European grassland swards.

Grass seeds include the coleoptile (shoot sheath), the scutellum, the radicula and the coleorrhiza (root sheath) (Figure 1.1). The scutellum is homologous to the leaf lamina of the cotyledon and the coleoptile to the leaf sheath of the cotyledon. Coleoptile and coleorrhiza are membranes that protect shoot and leaf meristems and the radicula, respectively, during the first step of the germination process when these fragile organs have to find their way through the soil. When a seed germinates, it first produces the radicula (the first root) that quickly absorbs water and nutrients. The coleoptile is then pushed upward and elongates to reach the soil surface, where the first leaf emerges from it. Side roots are produced quickly around the primary root. When the first three leaves are deployed, a bulge appears just above the seed and below the soil surface. It is the tillering plateau from which all secondary roots and aerial tillers are produced. Primary roots and seeds disappear afterwards.

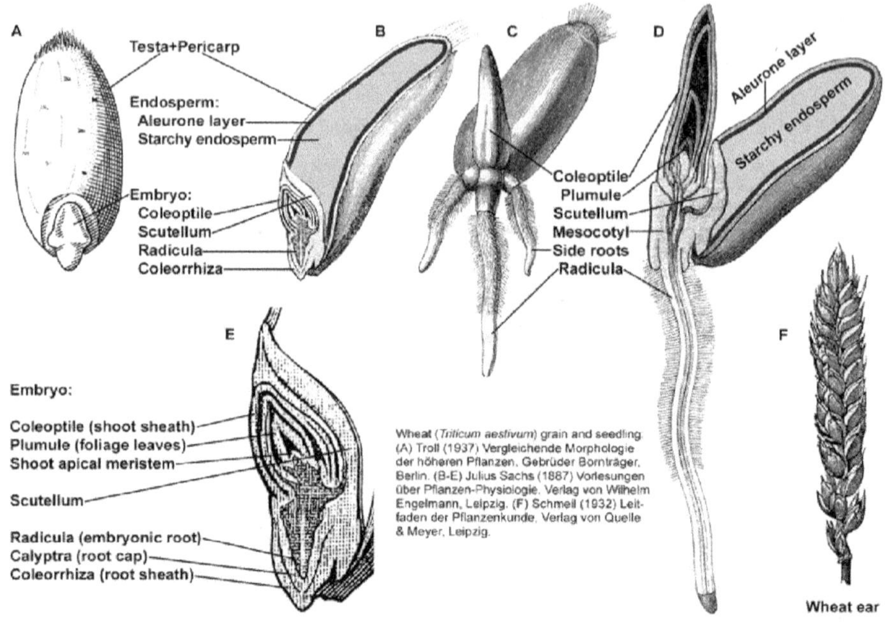

Figure 1.1. Seed anatomy. © The Seed Biology Place, Gerhard Leubner, http://www. seedbiology.de, 2019.

Grass leaves comprise three sections. The first is the upper part of the leaf, the lamina or blade, which is the most visible because it usually detaches from the stem). The second is the lower part of the leaf, the sheath, which encircles the stem (Figures 1.2 and 1.3). Finally, at the intersection of the lamina and the sheath, there can be two organs: the auricle and the ligule. Auricles are often claw-like appendages which tend to clasp the sheath. The ligule is an extension of the sheath at the base of the lamina. Its axis is parallel to the stem and thus perpendicular to the lamina. It may prevent water penetration between the sheath and the stem which could cause stem rotting. Tillers are the equivalent of branches in woody species. A tiller appears at the internal base of a leaf sheath. Each tiller can again produce leaves and new tillers at the base of these leaves. Tillering is thus theoretically exponential, but it is limited by light, nutrient and water resource availability. After germination, when four leaves are visible on the main tiller, there is a moment when two secondary tillers appear at the base of the first two leaves. The full **tillering phase** is then reached.

During the vegetative phase described above, all meristems are located just below or just above soil surface, keeping them well protected from herbivore teeth. This gives grasses a unique capacity to regrow quickly after defoliation compared to many dicotyledons. Grass plant species are thus well adapted to herbivore presence and activity. In fact, the two plant and animal species groups co-evolved and depend on each other. Most dicotyledon species are not especially well adapted to grazing because most of their meristem is located well above the soil surface. Because these plants can be easily destroyed by grazing animals, most swards grazed by herbivores are dominated by grasses. In a way, herbivores 'weed' herbaceous swards by reducing the proportion of dicotyledon species. In return, grasses produce more nutritious leaves to reward herbivores for this service. Unlike many dicotyledon species, grasses do not try to escape from grazing action by producing toxic compounds or mechanical defence organs; on the contrary, they encourage herbivores to graze them.

Stems can be produced on each tiller. A stem bears leaves and an inflorescence on its top. Some species requires a period of frost (vernalisation) to induce the reproductive phase and stem elongation. Others can enter into the reproductive phase soon after the tillering phase. These stems emerge from the tillering plateau where the roots and aerial organs meet. They are made of nodes on which leaves appear, and internodes, the space between two nodes. Internodes are initially very short. The first internodes start growing and push the stem upwards a bit like a radio antenna. In grass, the base elongates first. This is the **internode elongation phase**. When the first internode is about to reach its full length, the second internode, located above the first one, also begins to elongate. This is the **stem elongation phase**. The process continues with each following internode until the last leaf – called the flag leaf – emerges from the sheath of the previous leaf. Soon after, the inflorescence also emerges. This is the **heading phase**. When the inflorescence is completely deployed, etamins grow out from the flowers. This is the **flowering phase**. When seeds are formed, the plant has reached the **fructification (seed forming) phase**.

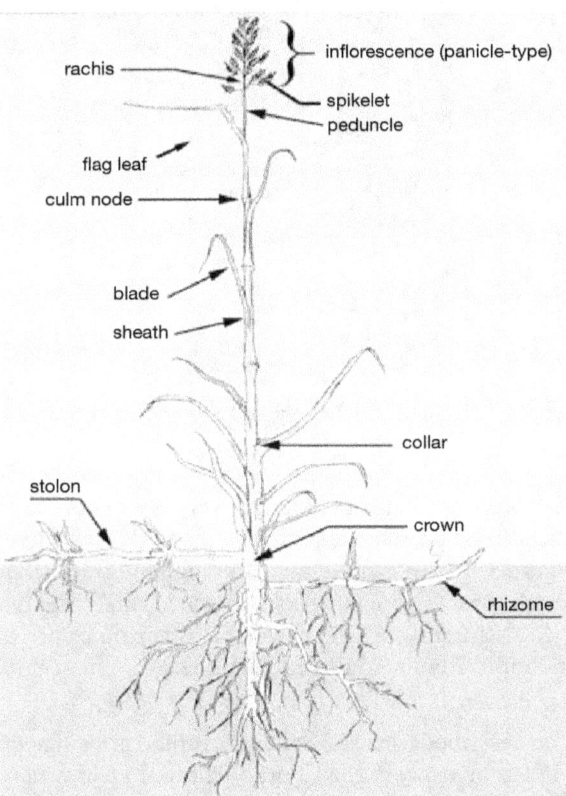

Figure 1.2. Grass anatomy (Oregon State University, Forage Information System, 2019).

Grass inflorescences can be spikes or panicles. Spikes are contracted inflorescence as in wheat (*Triticum aestivum*) or ryegrass (*Lolium* spp.). Spike ramifications are very short. Panicle ramifications are much more developed and visible, such as in oats (*Avena sativa*), fescues (*Festuca* spp.) or meadow grasses (*Poa* spp.). Inflorescences are made of spikelets that can contain one or several flowers. Each flower can produce a seed.

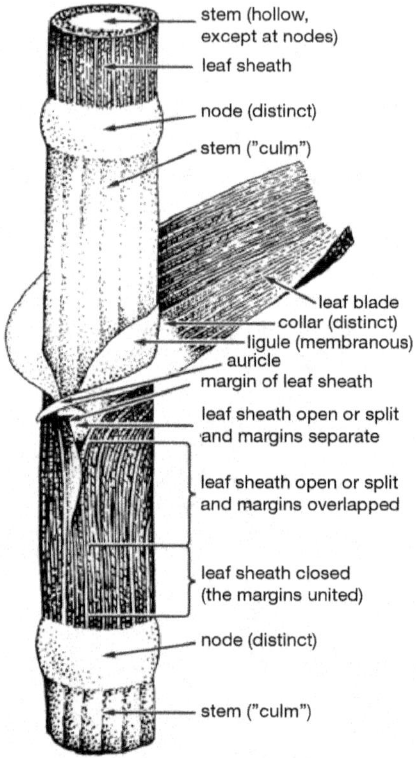

Figure 1.3. Leaf characteristics (Ministry of Agriculture, Food and Rural Affaires, Ontario, 2006).

Most grasses make compact tiller tufts, such as ryegrasses (*Lolium* spp.) or cocksfoot (*Dactylis glomerata*). Some produce creeping stems that grow above the soil surface, called stolons (Figure 1.2). An example is creeping bentgrass (*Agrostis stolonifera*). Others, such as couch grass (*Elytrigia repens*), produce rhizomes that are stems growing horizontally below the soil surface. Stolons and rhizomes are efficient means of vegetative reproduction. Plants equipped with these organs can easily spread in a sward.

Grass growth can be described for a single uninterrupted growing period or for several growing periods. It can be expressed as a production or yield which is the amount of biomass per surface unit (usually given as kg of dry matter (DM)/ha) or as a growth rate equal to the production per time unit (usually given as kg DM/ha x day).

Figure 1.4 shows an example of the evolution of dry matter production (Y axis) over time (X axis) during an uninterrupted grass growth cycle. Four phases can be distinguished in this growth cycle. At the beginning of the cycle, such as at the end of winter or after seed germination, grass growth is very slow (phase I). After 10 days on the figure, growth becomes very fast (phase II). This period lasts for 20 days on the figure (from day 10 to day 30). It is extremely fast between the 10th and the 20th day, then the growth curve passes through an inflexion point and it slows down slightly. In phase III, growth is almost nil and production reaches a maximum. After the 50th day on the figure, growth is negative because senescence exceeds growth and biomass decreases (phase IV).

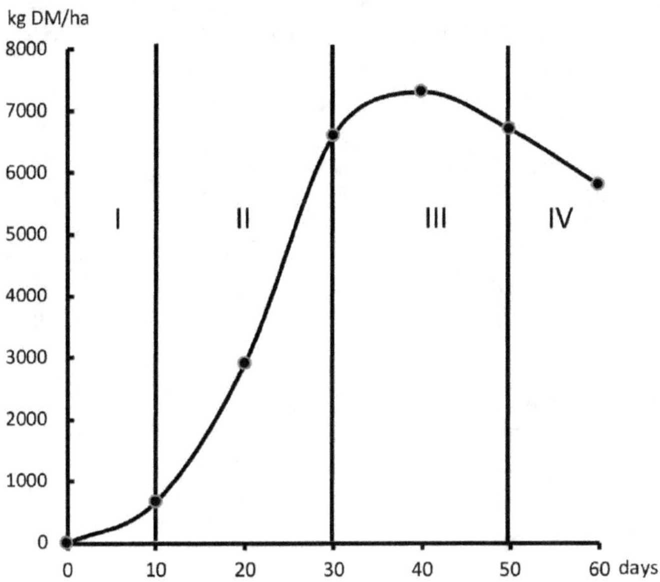

Figure 1.4. Evolution of biomass over time during an uninterrupted grass growth cycle (Peeters and Wezel, 2018).

This curve is exponential during phase I and the first part of phase II. It can be described as a Mitscherlich curve until the middle of phase III. The four phases of Figure 1.4 are common to many living beings. They correspond to juvenile, youth, adult and senile phases.

At the beginning of the growing period, plants have to mobilise their nutrient resources that can be either stored in seeds or in plant organs (stem base and roots). These resources, notably carbohydrates, are used for building new photosynthetic organs, with leaves in priority. In the beginning of phase II, plants will have produced a leaf area that can absorb enough solar energy to become independent from nutrient storage. At the inflexion point of phase II, plants will have produced so much carbohydrates that they are able to replace all nutrient resources consumed in phase I. A grass sward should thus not be defoliated by grazing or cutting before this point or the plants will be weakened, especially if the same process happens in the following growth cycles. This could eventually lead to plant death. In the second part of phase II, plant production is still increasing fast. By phase III, it is clear that the best cutting time is exceeded. There is no further biomass accumulation. In phase IV, part of plant biomass is lost because of leaf and stem senescence. These old organs lose weight and their nutrients are transferred to younger organs. They become pale green, yellow and finally brown. Eventually they detach from their plant and fall off, where they are recycled by soil organisms – mainly earthworms, bacteria and fungi.

The regrowth cycle initially relies on the energy reserves that plants store as carbohydrates in the basal stems. Immediately after grazing, plants rely on these to provide energy for regrowth until the first new leaf is produced (Figure 1.5). With the first new leaf, photosynthesis then becomes the main energy source for the growth of subsequent leaves, as well as replenishing carbohydrate reserve stores. Reserves are restored when plants have produced three new leaves.

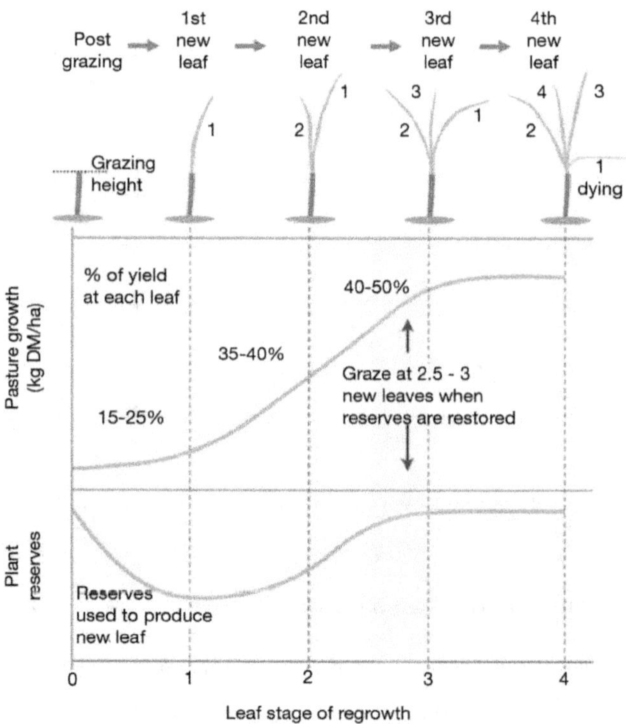

Figure 1.5. Regrowth and energy reserves in perennial ryegrass (Barenbrug Agriseeds, 2019).

Figure 1.6 presents a similar yield evolution curve. It also describes the evolution of physiological stages. In phase I, plants are in the vegetative phase. Above ground, they produce only leaves. In the first part of phase II (before the inflexion point), they begin to produce stems. This is the internode elongation stage, which is quickly followed by the stem elongation stage. Early heading corresponds to phase III and seed forming to phase IV.

Figure 1.6 also shows the evolution of forage quality in parallel to the evolution of grassland yield. Grass quality covers several parameters among which the most important are digestibility, net energy, protein, carbohydrate (soluble carbohydrate, cellulose and hemicellulose), lignin, and macro- and micro-element content. Good forage for demanding livestock types is highly digestible and contains high levels of protein, energy and minerals and low lignin levels. When plants grow, the proportions between cell content and cell walls decreases. Since cell content contains more nutrients than cell walls, quality decreases. During grass growth and development, the proportion of leaves decreases to the benefit of stems. Stems are less digestible than leaves because their lignin content is higher. Forage quality therefore evolves opposite to yield: when yield increases, forage quality decreases. A compromise must be made between forage quantity and quality for feeding livestock. This compromise is reached in spring at about the internode elongation stage or the beginning of stem elongation. When a sward is grazed at these stages, many stems are also destroyed and regrowth is leafy in some grasses and thus more nutritious. In summer and autumn, these grass species produce only leaves. The ideal grazing stage is thus vegetative.

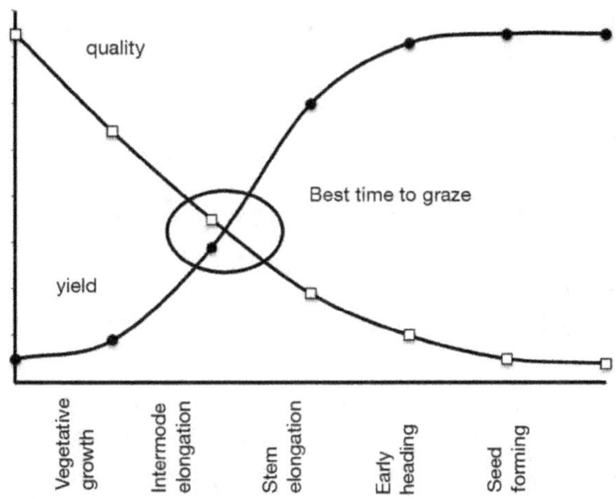

Figure 1.6. Evolution of dry matter production and forage over time during an uninterrupted grass growth cycle (Peeters and Wezel, 2018).

The production equation can be derived for calculating growth rate. If growth rate is calculated for the data of Figure 1.6, the growth rate curve has a clock shape with a maximum at the inflexion point of phase II. This means that growth rate accelerates from the beginning of phase I and decreases after the inflexion point to reach 0 in phase III and become negative in phase IV.

Figure 1.7 shows a sward growth rate evolution between March and November in an Atlantic climate of north-western Europe. This camel-back curve has two maxima. The first and the most important one is noted in May when the growth rate can reach 160 to 180 kg of DM produced per ha and per day. At that period of the year, in a single week, up to one ton of DM can be accumulated per ha. May is the period of the elongation phase. It is the middle of the reproductive phase of most grass species.

When most stems are destroyed by successive defoliations, the growth rate goes down. This decline is reinforced by summer water deficits. The curve then reaches its minimum point, usually in August. After summer drought, autumn rains stimulate nitrification in grassland soils. When these two combined factors of water and nitrogen availability revive plant growth. In September, a second but smaller maximum is reached. However, temperatures and sunlight are soon limited and the growth rate declines again. In continental climates, the growth rate is totally stopped in winter. In hyper-Atlantic climates such as those of Ireland, south-east England or Brittany, grass growth never totally stops in winter, but it is of course very low.

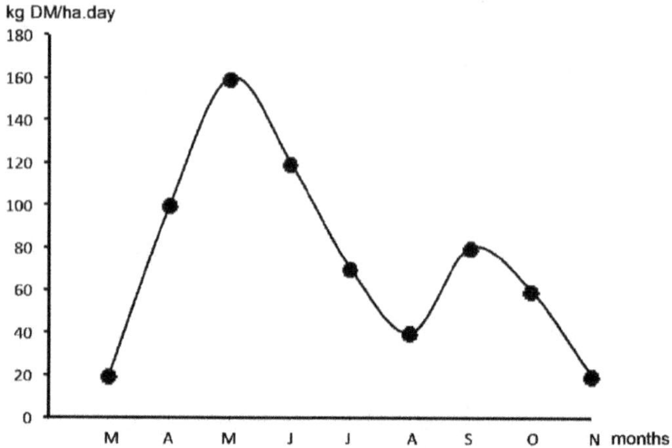

Figure 1.7. Example of an evolution of sward growth rate over a full growing period from March to November in an Atlantic climate of north-west Europe (Peeters and Wezel, 2018).

Grassland vegetation is not limited to grasses alone. Permanent grasslands can also host legumes and other forage plants. This last category includes non-leguminous dicotyledons (mostly *Apiaceae, Asteraceae, Caryophyllaceae Lamiaceae, Plantaginaceae, Ranunculaceae, Rosaceae* and *Scrophulariaceae*) and grass-like plants (e.g. sedge (*Carex* spp.) and rushes (*Juncus* spp.)). Even temporary grasslands can exhibit a certain plant diversity. Grasses and legumes are usually dominant, but plantain and chicory are more and more frequently used as forage plants and several spontaneous species may appear in the sown sward.

Non-leguminous dicotyledons have long been considered as weeds and chemically destroyed. Many are currently recognised as forage plants that can bring nutrients and secondary compounds that are beneficial to livestock. They may also increase forage intake by animals. Common dandelion (*Taraxacum* spp.) is an example of these valuable forage species. However, in permanent grasslands, a balance in the proportion of grasses, legumes and other species should be sought. There is no optimum, universal balance but the following proportions can be cited as one possible example: 60% grasses, 30% legumes and 10% other forage plants. A legume content of 30% to 50% is recommended for increasing livestock intake according to different studies.

▸▸ Use of grasslands

A. Peeters and N. Nilsdotter-Linde

Permanent grasslands often occupy priority plots that are considered regionally as marginal, i.e. plots that cannot or cannot easily be used for annual crops. Soils may be superficial, too stony, too wet, too dry or nutrient poor. These plots may also be located on steep slopes or in remote places in relation to farm buildings. Permanent grasslands are often grazed in regions where grazing is traditional.

They may also be placed near dairy cow sheds, even if these plots are located on very good, deep and nutrient-rich soils. Long moves between milking places and grazing land should be limited as much as possible for dairy cows.

In intensive dairy systems, when the grazing area near farm buildings is not limited, permanent grassland plots can be alternatively used for grazing and cutting. In this case, cuts are mainly conserved as silage. Alternating in this way between grazing and cutting can be interesting for several reasons. Cutting at the right time can help control perennial weeds by preventing seed formation of docks (*Rumex* spp.), thistle (*Cirsium* spp.) and nettle (*Urtica* spp.) for instance. Cuts are followed by urine and dung-free regrowth that is more palatable for grazing animals. This regrowth is clean from livestock internal parasites. Refuse from the previous grazing periods (mainly stems) are harvested by cuts which generate a fresh leafy sward afterwards. Cuts are taken after a longer regrowth period than grazed grass. This is favourable to the restoration of nutrient storage in the shoot bases and roots of grassland plants. Meanwhile, grazing makes swards denser, maintains short sward heights that are favourable to the persistence of white clover and some grasses, and fertilises plots with urine and dung.

In beef systems and in the case of heifers, permanent grassland plots are mainly grazed and rarely cut. They can be located at some distance from farm buildings since animals are not brought back every day to a barn. They can stay in the same location for several months, moving or not from one plot to another.

In regions where grazing is not traditional, where animals are kept indoors the whole year, permanent grasslands can be exclusively cut for harvesting fresh grass, or for silage or hay. However, this type of management causes sward quality to degrade over time, especially in the case of late cuts. Forage exports must also be compensated by organic or mineral fertilisation to avoid soil nutrient depletion. This system is also more expensive compared to grazing systems where animals harvest grass themselves, while in zero grazing systems and with conserved grass, forage must be cut, possibly dried out, transported by mechanical means and distributed fresh.

Permanent grassland vegetation is the result of the many interactions between soil, climate and farmer management. It is thus highly variable throughout Europe. Permanent grasslands can be agriculturally-improved, semi-natural, natural or no longer used for production.

Temporary grasslands can be either grazed, cut or mixed use. There is, however, a clear trend to use them for winter forage production in addition to grazed or mixed-used permanent grasslands. They are often included in crop rotations with annual crops. Temporary grasslands can be sown with pure grass, pure legume or grass/ legume mixtures. Innovative mixtures based on plantain, chicory or white and red clovers, for instance, are developing for specific uses. Pure grass mixtures are usually fertilised. They can include one or several species of the following grasses: perennial, Italian and hybrid ryegrasses, tall fescue, meadow fescue, timothy or cocksfoot. Ryegrasses are usually dominant in Atlantic climates. Pure legume mixtures are rarer and often located in southern Europe. Pure lucerne or pure sainfoin can be noted in some southern areas. Grass/legume mixtures are very common. Examples of simple mixtures are cocksfoot/lucerne, ryegrass/red or white clover, cocksfoot/red clover, timothy/red clover. More complex mixtures can include perennial ryegrass/meadow fescue/timothy/red and white clovers. Many solutions are possible.

Temporary grasslands can also be grazed and occasionally cut. A typical situation is often encountered in north-western Europe, where a temporary sward is resown every four to five years with perennial ryegrass, which may or may not be mixed with other grasses and white clover, or which may be regularly oversown. It is thus not included in a crop rotation. It may, however, be established on very good, deep and nutrient-rich soils. This sward type is often grazed by dairy cows, and occasionally by suckler cows.

Mainly cut temporary grasslands can be used for extending the grazing season along with permanent grasslands. Red clover/grass mixtures start growing earlier than permanent grasslands in spring. They can be grazed at that time for about two weeks with electric fencing. The animals are then moved to permanent grassland plots where they graze until around mid-October, while temporary grasslands may be cut up to three times for silage making. In north-western Europe, when permanent grasslands stop growing in autumn, animals can again graze temporary grasslands through the end of December or January depending on the location and weather conditions. This technique can dramatically reduce winter feeding costs by reducing the housing period from six to three months. However, grazing on wet soils in early spring and late autumn requires caution to not damage the sward and soil structure. The further north, the more important it is to also consider the effect of late autumn grazing on winter survival. Late defoliation decreases the reserves of carbohydrates needed for winter survival if there is no time for regrowth before winter arrives.

Exclusively grazed grasslands can be permanent or temporary. However, they are mainly permanent and grazed by suckler cows or heifers. They can also exist in a dairy cow system when grazing area is limited around farm buildings. Dairy cow plots are exclusively grazed in this case. These plots may be permanent or temporary grasslands.

In the past, in traditional extensive systems, specialised cutting grasslands existed everywhere. They were cut once for hay in summer and the small regrowth was then usually grazed in September. They were located on a range of soils that were very acidic, calcareous or wet. Their vegetation was among the most diverse of all European terrestrial vegetation. They were progressively abandoned and are now threatened and a topic of nature conservation concern.

In conventional farming, **exclusively cut grasslands** are mainly temporary. An exclusive cutting regime has indeed a negative influence on sward quality. In two- to three-year temporary grasslands, for instance, this effect is acceptable while in permanent grasslands it leads to strong sward degradation after a couple of years. However, exclusively cut permanent grasslands can still be found, particularly in large wet valleys where soils are too wet in spring to be grazed. The first cut is then taken later when soils are better drained and allow for tractor traffic. If these plots are also far from farm buildings, the later regrowth can also be harvested by cutting. This situation is common in eastern Poland. Foxtail (*Alopecurus pratensis*) or couch grass (*Elytrigia repens*) are often the dominant grasses in these conditions.

Mixed-used grasslands are very common either in permanent or temporary grasslands. Most extensive systems combine a hay cut with several grazing periods. Most intensive systems, usually associated with dairy systems, may combine several short grazing periods with one or two silage cuts. As explained above, alternating between grazing and cutting is a very beneficial system for both grassland plants and grazing animals.

▶▶ Grassland species and sward assessment

G. Peratoner, M. Komainda and J. Isselstein

Factors affecting the botanical composition

Grasslands provide ecosystem functions, such as carbon sequestration, nitrogen fixation (if legumes grow in the sward), water quality preservation, biodiversity conservation, positive effects if integrated into a crop rotation (prevention of weed infestation in arable crops), cultural aspects and the provision of forage for ruminant livestock production. Due to a wide range of climates (sub-tropic to sub-arctic, from mountainous to the lowlands), grasslands appear in diverse botanical compositions. The number of species is mainly mediated by the management intensity and can range between a few species in temporary grasslands and up to very high numbers (e.g. about 70) in extensively managed grasslands. Moreover, the magnitude of these compositions is driven by the management intensity (fertilisation, defoliation frequency) and all other agronomical measures (e.g. cut vs. grazed, over- and re-sowing) at a specific site. Each species has its own requirements (functional traits) with respect to the management and site conditions and a sustainable sward is only as good as it is managed, because the grassland sward is a result of several interacting factors (Figure 1.8). In fact, all of these factors directly or indirectly affect competition between species, which is an underlying main driver of changes of the botanical composition over time. Site conditions are the basis and preconditions of productivity under the given environment, as they affect plant growth (climate, weather, soil properties) and the mechanisation potential (topography). Usually, they cannot be changed by the farmer. Utilisation form, fertilisation and sward care depend on the farmer's agronomical choices and represent different management aspects. There are farms that predominantly cut their grass, others that graze it and some that do both. Intensity refers to the number of cuts per year or the grazing intervals on one site. In cold climates, the timing of the first and the last cut influences the winter survival and the botanical composition of the sward. Nutrients affect soil fertility and thus productivity up to a certain point. With increasing nutrient availability, species taking most advantage of fertilisation tend to become dominant. Over-fertilisation does not further increase forage yield and results in species-poor swards with little biological and agronomical value. With sward care, the operations periodically carried out to maintain good conditions of the swards are addressed (i.e. resowing or rolling in spring). Concerning sward renovation, the composition of the seed mixture greatly affects the botanical composition of young swards, while grassland management becomes more and more important over time. Consequently, the sward botanical composition is a result of complex interactions.

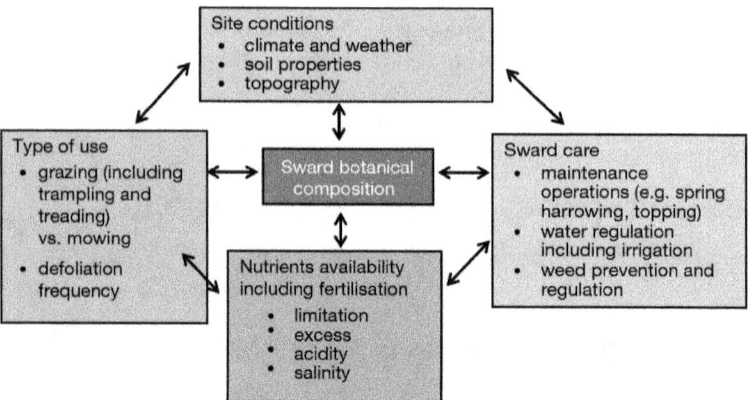

Figure 1.8. Factors affecting the botanical composition of the sward (adapted from Voigtländer and Jacob, 1987).

Grasses, legumes and forbs

The most important functional groups of species in a grassland sward are represented by grasses, legumes and forbs (herbaceous, mainly dicotyledonous species being neither grasses nor legumes), all of which are more or less likely to occur in forage production. Examples of common forbs in grassland swards for dairy production are dandelion (*Taraxacum officinale*), plantain (*Plantago lanceolata*) or common chickory (*Cichorium intybus*). Shrubs, bushes and trees mostly appear at the edges of pastures, spontaneously in the swards or organised in agroforestry systems, and represent a feeding source for browsing animal species (i.e. sheep and goats).

Grasses are able to tiller, i.e. to generate secondary tillers from tiller buds. Tillering is stimulated by defoliation (mowing, grazing) and the position of the tiller buds determines the growth form of grasses. Grasses with aboveground upright/erect growing tiller buds form tussocks, while grasses with belowground tiller buds (turf-forming) are particularly efficient in building dense swards (Table 1.1). Some species have a pronounced creeping habit due to specialised tiller buds enabling horizontal growth (stolons if growing aboveground and rhizomes if growing belowground) and are specialised in colonising bare ground or gaps in the sward. In meadows, grasses are more abundant at the first cut, as at this time they have entered the generative phase and there is a relevant contribution of their stems to the yield.

28

The opposite is true for legumes and forbs, which usually contribute relevant amounts in the regrowth. Since upright/erect grasses regenerate from the stubble after harvest, they are less adapted for intensive utilisation. Turf grasses regenerate from the remaining leaf area after harvest or grazing, since the growing point remains at the bottom of the sward when defoliated. Therefore, they require/tolerate frequent harvests or grazing. Otherwise, the oldest (lower) leaf decays due to unfavourable light conditions.

Table 1.1. Main differences between tussock-forming and turf-forming grass species. This information is general and there are exceptions depending on the single species and location.

Type	Tussock grass	Turf grass
Reproduction	Mainly generative	Mainly vegetative
Tolerance to frequent defoliation	Low	High
Tolerance to grazing	Low	High
Sward density	Low	High
Deterioration of forage quality with advance in phenological development	High	Low
Examples	*Dactylis glomerata* (Cocksfoot), *Festuca pratensis* (Meadow fescue), *Phleum pratense* (Timothy), *Phalaris arundinacea* (Reed canary grass)	*Lolium perenne* (Perennial ryegrass), *Poa pratensis* (Smooth-stalked meadowgrass), *Festuca rubra* (Red fescue)

Another categorisation of forage grasses is based on whether a cold period (vernalisation) before the generative stage after the spring growth is needed (leaf grasses) or not (stem grasses). Examples of leaf grasses are cocksfoot (*Dactylis glomerata*), meadow fescue (*Festuca pratensis*), smooth-stalked meadow grass (*Poa pratensis*) and red fescue (*Festuca rubra*). Reed canary grass (*Phalaris arundinacea*) is a typical stem grass and timothy (*Phleum pratense*) is more like a stem grass than a leaf grass. The stem grasses are more vulnerable to frequent defoliation compared to leaf grasses.

The advantage of **legumes** in the sward is their ability to fix atmospheric nitrogen thanks to symbiosis with bacteria (*Rhizobium* spp.) and to contribute to high leaf proportions of good quality, i.e. high protein concentration and low fibre content. They thus allow to reduce nitrogen fertilisation without decreasing production, and to increase forage quality.

Forbs can highly vary in their morphology, but they usually contain high amounts of macro nutrients, secondary metabolites (both favourable and unfavourable, depending on the species) and have favourable digestibility because of their high proportion of leaves. Forbs rich in rough stems (i.e. *Apiaceae*) are usually less favourable in terms of forage quality. It is a rule of thumb that a higher forage intake can be achieved if swards contain up to 20% of non-toxic forbs. If their proportion exceeds certain species-specific thresholds, forbs can be detrimental for the sward density, resulting in smaller yields, higher permeability to weed invasion and disadvantages with respect to forage conservation (i.e. crumble losses in case of haymaking). Moreover, the most relevant poisonous species belong to this group.

Characteristics of the main grassland species

Knowledge of grassland species (both their determination and an understanding of their ecological and agronomical traits) is a prerequisite to understand what is happening in the sward and to adapt management in a timely manner. Otherwise, no appropriate measure for assessment of the actual sward quality is possible with respect to the targeted aims (e.g. high forage quality). To this end, some of the relevant characteristics of the most frequently occurring species in the temperate European grasslands are summarised in two tables: Table 1.2 provides basic information about species morphology, while Table 1.3 focuses on the agronomical traits of the same species. Some facultative weeds (species undesirable, if occurring in high proportions) and absolute weeds (species undesirable or noxious regardless of their abundance) are included as well.

Determination of grassland species

Only a short description of the determination of grasses will be presented here. Because defoliation often occurs before flowering time of all species in productive grasslands, it is especially important to master the determination at the vegetative stage. Good keys are available in the respective national languages (for example, English: Hubbard, 1992; Frame and Laidlaw, 2014; German: Klapp and Opitz von Boberfeld, 2011; Klapp and Opitz von Boberfeld, 2004, Dietl *et al.*, 1998; Dietl and Jorquera, 2003; Italian: Dietl *et al.*, 2003; Swedish: Westerlind *et al.*, 1997). However, the most relevant morphological traits for the determination of grasses can be used as follows:
– **Step 1:** Is the youngest (uppermost) leaf rolled or folded?
– **Step 2:** How is the texture of the leaf blade (lamina)?
– **Step 3:** How is the surface and respective bottom of each leaf formed? Is it rough or shiny?
– **Step 4:** How is the leaf base (auricle, ligule) formed?
– **Step 5:** If occurring at the time of the assessment: What does the inflorescence tell you about the species?

Table 1.2. Morphology and ecological characteristics of the main grassland species. The table is a compendium of numerous sources (among them Klapp and Opitz von Boberfeld, 2011; Dietl and Jorquera, 2003; Landolt *et al.*, 2010) and the author's own experience.

Botanical name	Common name	Growth form: 1=tussock, 2=turf	Youngest leaf: 1=rolled, 2=folded, 3=bristle-like	Texture of upper side of leaf blade: 1=ribbed, 2=smooth, 3=tramlined	Bottomside of leaf blade: 0=not shiny, 1=shiny	Auricles	Ligule	Hairiness: 1=yes; 0=no	Inflorescens and spikelet	Vegetative reproduction	If vegetative: 1=rhizome, 2=stolon	Winter hardiness	Suitable for wet sites	Suitable for dry sites	Soil reaction (RZ)
Grasses															
Agrostis capillaris	Common bent	1	1	1	0	no	yes	0	panicle	no		+++	+	-	2
Agrostis stolonifera	Creeping bentgrass	2	1	1	0	no	yes	0	panicle	yes	2	++	+	+	4
Alopecurus pratensis	Meadow foxtail	1	1	1	0	no	yes	0	spike	yes	1	+++	++	-	3
Anthoxanthum odoratum	Sweet vernal grass	1	1	1, 2	0	yes	yes	1	panicle	no		++	+	+	2
Arrhenatherum elatius	Tall oat-grass	1	1	1	0	no	yes	1	panicle	no		+	+	+++	4
Dactylis glomerata	Cocksfoot, Orchardgrass	1	2	2	0	no	yes	0	panicle	no		+++	+	++	4
Deschampsia cespitosa	Tufted hairgrass	1	1	1	0	no	yes	0	panicle	no		+++	+	+	3
Elymus repens *Elytrigia repens*	Couch grass	2	1	1, 2	0	yes	yes	1	spike	yes	1	++	++	++	4
Festuca arundinacea	Tall fescue	1	1	1	1	yes	yes	0	raceme	no		++	-	+++	4
Festuca pratensis	Meadow fescue	1	1	1	1	yes	yes	0	raceme	no		+++	+	-	4
Festuca rubra	Red fescue	1	3	1	1	no	yes	0	raceme	yes	2	+++	+	++	3
Holcus lanatus	Fog grass	1	1	2	0	no	yes	1	panicle	no		+	+	+	3
Lolium perenne	Perennial ryegrass	1,2	2	1	1	yes	yes	0	panicle	no		+	+	-	4
Phleum pratense	Timothy	1	1	2	0	no	yes	0	spike	no		+++	+	-	3
Poa pratensis	Smooth-stalked meadowgrass	2	2	2, 3	1	no	no	0	panicle	yes	1	++	+	-	3
Poa trivialis	Rough-stalked meadowgrass	1,2	2	2, 3	1	no	yes	0	panicle	yes	2	++	+	+	3
Trisetum flavescens	Gold oat grass	1	1	1	0	no	yes	1	panicle	no		++	+	+	3
Legumes															
Lotus corniculatus	Birdsfoot trefoil									no		++	+	++	4

Botanical name	Common name	Growth form: 1=tussock, 2=turf	Youngest leaf: 1=rolled, 2=folded, 3=bristle-like	Texture of upper side of leaf blade: 1=ribbed, 2=smooth, 3=tramlined	Bottomside of leaf blade: 0=not shiny, 1=shiny	Auricles	Ligule	Hairiness: 1=yes; 0=no	Inflorescens and spikelet	Vegetative reproduction	If vegetative: 1=rhizome, 2=stolon	Winter hardiness	Suitable for wet sites	Suitable for dry sites	Soil reaction (RZ)
Medicago lupulina	Black medic									no		++	-	+	4
Medicago sativa	Alfalfa									no		+	-	+++	4
Onobrychis viciifolia	Sainfoin									no		-	-	+	4
Trifolium hybridum	Alsike clover									no		++	++	-	4
Trifolium pratense	Red clover									no		++	+	+	4
Trifolium repens	White clover									yes	2	++	+	+	4
Forbs															
Achillea millefolium	Yarrow									no		++	+	+	
Anthriscus sylvestris	Cow parsley									no		++	+	-	
Carum carvi	Caraway									no		++	+	-	
Cichorium intybus	Common Chickory									no			+	+	8
Cirsium spp.	Thistle									no		+	+	-	7
Daucus carota	Wild carrot									no		+	+	+	
Geranium pratense	Meadow crane's-bill									no		++	+	-	8
Heracleum sphondylium	Common hogweed									no		++	-	++	
Plantago lanceolata	Ribwort plantain									no		++			
Ranunculus acris	Meadow buttercup									no		++	++	-	
Ranunculus repens	Creeping buttercup									yes	2	++	++	--	
Rumex acetosa	Common sorrel									no		++			
Rumex crispus	Curly dock									no		++	++	--	
Rumex obtusifolius	Broad-leaved dock									no		++	+	-	
Senecio spp.	Ragwort												+	+	7
Taraxacum officinale	Dandelion									no		++	+	-	

RZ refers to the soil reaction requirement ranging from 1 (extremely acid) to RZ 5 (alkaline, high soil pH).

Table 1.3. Agronomical traits of the main grassland species. The table is a compendium of numerous sources (among them Klapp and Opitz von Boberfeld, 2011; Dietl and Jorquera, 2003; Landolt *et al.*, 2010) and the author's own experience.

Botanical name	Common name	Properties								Suitable for								
		Competition at establishment	Earliness at booting	Proneness to dominance under favourable conditions	Yield potential	Forage quality	Deterioration of forage quality with phenological advance	Palatability	Overall forage value (WZ)	Mowing	Grazing	Cut frequency	Growth reaction to N-input	Field drying	Barn drying (drying facility)	Silage	Temporary grassland	Threshold to weed status (yield proportion in %)
Grasses																		
Agrostis capillaris	Common bent		-	+	+	+	+	++	5	++	+	1-2	+	++	++	-	-	-
Agrostis stolonifera	Creeping bentgrass	+	++	+	+			+	5	++	-	1-2	+	+	++	+	-	-
Alopecurus pratensis	Meadow foxtail	+	++++	+++	+++	++	++	++	7	++	+	2-4	+++	+	++	+++	-	-
Anthoxanthum odoratum	Sweet vernal grass		+++	-	-	+		+	3	++		1-3	+				-	-
Arrhenatherum elatius	Tall oat-grass	++	++	++	++	+	++	++	7	++	-	1-2	++	++	++	+	-	-
Dactylis glomerata	Cocksfoot, Orchardgrass	++	++	++	+++	+	+	++	7	++	+	1-4	++	++	++	+++	++	-
Deschampsia cespitosa	Tufted hairgrass		+	++	++	-	+++	-	3	+	-	1-2	+	++			-	10
Elymus repens	Couch grass	+	++	+++	++	-	+	+	4	++	-	1-3	+++	++	++	++	-	20
Festuca arundinacea	Tall fescue	-	++	++	++	+	++	+	4	++	±	3-4	++	++	++	++	-	-
Festuca pratensis	Meadow fescue	+	+	+	+	++	-	+++	8	++	+	2-4	+++	++	++	+++	+	-
Festuca rubra	Red fescue	-	++	++	+	+	+	++	5	++	+	1-3	+	++	++	+	-	-
Holcus lanatus	Fog grass		++	-	-	-		-	4	+		1-3	+	++	++	++	-	15
Lolium perenne	Perennial ryegrass	+++	+++	+++	+++	+++	-	+++	8	++	+++	2-5	+++	+	++	+++	++	-
Phleum pratense	Timothy	+	-	+	+	++	+	++	8	++	+	1-4	+++	++	++	+++	++	-
Poa pratensis	Smooth-stalked meadowgrass	-	++	++	++	+++	-	+++	8	++	+++	3-5	++	++	++	++	+	-

Botanical name	Common name	Properties							Suitable for									
		Competition at establishment	Earliness at booting	Proneness to dominance under favourable conditions	Yield potential	Forage quality	Deterioration of forage quality with phenological advance	Palatability	Overall forage value (WZ)	Mowing	Grazing	Cut frequency	Growth reaction to N-input	Field drying	Barn drying (drying facility)	Silage	Temporary grassland	Threshold to weed status (yield proportion in %)
Poa trivialis	Rough-stalked meadowgrass	+	++	++	-	++	+	-	4	++	-	3-5	++	++	++	+	-	15
Trisetum flavescens	Gold oat grass	+	+	++	++	++	+	++	4	++	-	1-3	+	++	++	-	-	-
Legumes																		
Lotus corniculatus	Birdsfoot trefoil	-	++	++	+	++	-	++	7	++	+	1-3	-	++	+++	+	-	-
Medicago lupulina	Black medic	-	++	+	-	++	-	++	7	++	+	1-3	-	+	++	+	-	-
Medicago sativa	Alfalfa	++	++	+++	+++	+++	-	++	8	++	-	2-5	-	-	++	+	++	-
Onobrychis viciifolia	Sainfoin	+		++	+	++	-	++	7		-		-	+	++	+	+	-
Trifolium hybridum	Alsike clover	-	+++	+	+	++	-	++	7	++	-	1-3	-	+	+++	+	+	-
Trifolium pratense	Red clover	+	++	++	++	++	-	++	7	++	+	2-5	-	++	+++	+	++	-
Trifolium repens	White clover	++	+++	+++	+	+++	-	++	8	++	+++	1-5	-	++	++	+	++	-
Forbs																		
Achillea millefolium	Yarrow		+	-	+	-	+	5	+	+	1-4	-	+	++	+	-	10	
Anthriscus sylvestris	Cow parsley		+++	+	+	+	++	4	++	-	1-3	++	-	+	+	-	15	
Carum carvi	Caraway		+	+	++	-	++	5	++	+	1-3	+	+	++	+	-	20	
Cichorium intybus	Common Chickory		++	++	++	-	++		++		1-4			+	+	+	-	
Cirsium spp.	Thistle		++	+	-	+	-	0	-	-	1-2	+				-	2	
Daucus carota	Wild carrot		+	+	++	-	++	5	++	-	1-2	+	+	++	+	-	-	
Geranium pratense	Meadow crane's-bill		+++	+	-	+	-	2	++	-	1-2	++	-	+	+	-	15	

		Properties								Suitable for									
Botanical name	Common name	Competition at establishment	Earliness at booting	Proneness to dominance under favourable conditions	Yield potential	Forage quality	Deterioration of forage quality with phenological advance	Palatability	Overall forage value (WZ)	Mowing	Grazing	Cut frequency	Growth reaction to N-input	Field drying	Barn drying (drying facility)	Silage	Temporary grassland	Threshold to weed status (yield proportion in %)
Heracleum sphondylium	Common hogweed			+++	+	+	+	++	5	++	-	1-3	++	-	+	+	-	10
Plantago lanceolata	Ribwort plantain			+	+	+	-	++	5	++	+	1-3	+	+	++	+	-	15
Ranunculus acris	Meadow buttercup			++	+	-	+	-	-1	++	-	1-3	+	-	+	-	-	5
Ranunculus repens	Creeping buttercup			++	+	+	+	+	2	++	+	1-3	+	-	+	-	-	10
Rumex acetosa	Common sorrel			++	+	+	+	+	3	++	-	1-3	+	+	++	+	-	15
Rumex crispus	Curly dock			+++	++	-	++	+	1	++	-	1-4	++	-	+	+	-	1
Rumex obtusifolius	Broad-leaved dock			+++	++	-	++	+	1	++	-	1-4	++	-	+	+	-	1
Senecio spp.	Ragwort			++	+	--		-	-1	+	-						-	1
Taraxacum officinale	Dandelion			++	+	++	-	++	5	++	+	2-4	++	-	+	+	-	20

Rating ranges between +++ very high or highly suitable to – very low or unsuitable. Overall forage value (WZ) ranges from 8 (maximum value) to 0 (no value) and -1 (poisonous species).

Sward assessment

The regular and periodic assessment of the botanical composition of the plant stand represents a valuable management tool in grassland farming, as this allows farmers to recognise and interpret ongoing trends in the vegetation and properly adapt their management.

In order to get reliable information about the sward, a few rules must be followed:
− A surveyed area of about 50 square metres is usually sufficient in managed grasslands.
− If the surveyed grassland is not homogeneous, the choice of a representative area may be difficult. In this case more than one area should be assessed.
− The yield proportion of species or species groups usually changes over the season (i.e. in meadows, grasses are most abundant at the first cut, while the opposite is true for legumes and forbs); if assessments over years are to be compared, the assessment should be temporally and spatially homogenous.

The abundance of species can be described using different parameters (plant density, frequency, cover), but for agronomic objectives, the yield proportion of species is highly relevant as it describes the contribution of each species to the yield. The yield proportion is the relative proportion (in percent of weight) of the harvestable aboveground dry matter biomass of a certain species or a species group related to the total dry matter yield. For the assessment, several methods differing in accuracy, time consumption and necessary tools to perform the assessment are available, but for practical reasons a visual estimation results in acceptable accuracy, as long as the same observer with a minimum of training repeats the assessments over time.

In order to assess grassland biodiversity, a complete species list is necessary, implying that all occurring species are determined and assessed. For practical reasons of grassland management, however, a simplified system based on the yield proportion of species groups (grasses, legumes, forbs) can be used. To this end, graminoids (sedges, rushes) are included into the grass group. Depending on the yield proportion of these groups, the plant stand can be assigned to one of four groups (Figure 1.9): rich in grasses, balanced, rich in forbs or rich in legumes. Already with this simple tool, relevant information can be gained about grassland management (Table 1.4).

Figure 1.9. Key to determine the sward type based on the yield proportion of grasses, legumes and forbs. Source: adapted from webGRAS (https://webgras.civis.bz.it/#/start).

Table 1.4. Hints about grassland management that can be inferred based on a simplified sward assessment of mixed swards in permanent grassland.

Plant stand type	Rich in grasses	Balanced	Rich in forbs	Rich in legumes
Yield potential	++	+	+	+
Crude protein potential	-	+	+	++
Forage quality stability along phenological development at the first cut	-	+	+	++
Easiness of silage conservation	++	+	-	-
Drying speed	++	+	-[a]	+
Risk of crumbling losses	-	-	+	+
Risk of contamination with soil	-	-	+	+

++ high, + medium, - low
[a] mainly in the case of forbs with rough stems

The actual grassland sward is a result of complex interactions with respect to the soil, management, climate and care. If famers find that the present grassland conditions do not meet expectations, several options are available. The first is to reconsider the management approach. The species occurring in the sward can be used as indicators of what is happening. Management intensity, including defoliation frequency and fertilisation, should be verified and, if appropriate, changed or adapted. Sward care should be applied to promote an increasing proportion of the desired forage species. The introduction of valuable forage species or the increase of their yield proportion can be achieved by means of a periodic oversowing (no need for complete sward renewal) or, if there is an absence of a critical mass of valuable forage species, by means of resowing, which requires good knowledge.

▸▸ Weed management

A. Peeters and N. Nilsdotter-Linde

As described in Section 1.1, permanent grassland vegetation often includes grass, legumes and other species. This last category comprises non-leguminous dicotyledons and grass-like plants. Even temporary grasslands can exhibit a certain plant diversity. Grasses and legumes are usually dominant, but several spontaneous species may appear in the sown sward.

Non-leguminous dicotyledons have long been considered **weeds** that should be destroyed by herbicides or other means. Many are currently recognised as forage plants that can bring nutrients and secondary compounds that are beneficial to livestock. They may also increase forage intake by grazing animals. They can be called 'herbs'.

A **weed** can be simply defined as a plant that appears where it is not supposed to be. Real weeds are restricted to a few species. There are two categories: annual and perennial species.

Annual species include chickweed (*Stellaria media*), fat hen (*Chenopodium* spp.), common knotgrass and redshank (*Polygonum aviculare* and *Persicaria maculosa*), annual nettle (*Urtica urens*), and chamomiles (*Matricaria* spp.). They may compete excessively with forage species just after sowing of a new sward. This could lead to insufficient forage plant establishment. Even if forage plants are well established, the presence of some annual weeds, especially fat hen and chamomile, in the young sward can decrease the forage quality of the next harvest.

There are several techniques that reduce annual weed invasion risk. The following practices are the most important:
– careful sowing bed preparation
– choice of an adequate seed mixture
– choice of a favourable sowing period
– topping of the young sward before the first harvest

When there is a risk of weed invasion in the plot chosen for establishing a grass sward, a false or stale seed bed can be prepared several weeks before grassland sowing. This could be done in the summer after the cereal harvest or in the spring (see comment on sowing period choice below). The soil is well prepared as it would be for an actual grassland sowing. This stimulates weeds germination. Once weeds have massively germinated and reached the cotyledon or the first leaf stage, they are mechanically destroyed by the superficial passage of a harrow or other means. This operation can be repeated once or twice if necessary. Then forage seeds are sown in a clean seed bed.

Forage plant seeds are very small. They require a very well prepared and fine seed bed for ensuring a close contact between soil and seed. After sowing, rolling is often advised for further improving soil-seed contact. These measures ensure fast and regular germination, which increases the chances of good establishment and weed control.

The outcome of competition between young forage plants and weeds very much depends on the germination rate and establishment speed of forage grasses and legumes. There are significant differences in germination and tillering speed between grasses. Italian ryegrass germinates and establishes very fast; perennial ryegrass is fast; cocksfoot and tall fescue are slow; timothy is slow and a weak competitor. Pure timothy sowings are unsurprisingly often invaded by weeds. Large differences are also noted in legumes. Annual crimson clover germinates and establishes very fast; red clover is fast; lucerne is slow; white clover is slow and requires a long establishment phase during which its competitive ability is low. An Italian ryegrass/red clover mixture, for example, would cover the soil surface very quickly and is able to efficiently control weeds.

In contrast, sowing density is not as important for weed control. Above the recommended sowing density threshold, (e.g. 30 kg of perennial ryegrass seed per ha) no significant effect of increased density on weed control is noted.

The choice of the sowing period is extremely important. Autumn sowing (second half of August – beginning of September) is safer than spring sowing (April – May) in north-western Europe. For example, fat hen mainly establishes in late spring sowing. This tall and branched plant is extremely competitive, and its very fibrous stems are poorly digested. In autumn, sowing should not be overly delayed or germination could be too slow and young plants could even be destroyed by winter frost. Legumes are more demanding than grasses in terms of soil temperature. They should be sown before 15 September in north-western Europe and before 1 August in the Nordic countries. This date varies according to location and weather conditions.

The sowing date also very much depends on favourable soil humidity conditions. Sowing in dry conditions increases the risk of fat hen, common knotgrass and redshank invasion because these plants can germinate in conditions that are too dry for forage grass and legume seeds.

Even when the preceding measures are adopted, young swards can still sometimes contain too many weeds. The abundance of these weeds could be judged as too high for a good quality harvest. Topping is then an option for reducing their profusion. Annual weeds usually establish faster and grow faster than perennial forage plants. When weeds start their reproductive period and produce stems, these stems are often significantly taller than the average height of forage plants. The sward can then be cut below the average height of forage plant canopy. When conditions are favourable, most annual weeds are then destroyed or at least their proportion in the sward is significantly reduced. New weed seeds can sometimes germinate later, but the risk is much lower when there is a dense sward of the desirable species.

Perennial species weeds include docks (*Rumex obtusifolius* and *R. crispus*), creeping thistle (*Cirsium arvense*), stinging nettle (*Urtica dioica*), ragwort (*Senecio jacobea*) and couch grass (*Elytrigia repens*).

Moreover, creeping buttercup (*Ranunculus repens*), when too abundant in a sward, can also reduce yield and forage quality. This situation is encountered on poorly drained and wet soils where inadequate management leads to progressive grass disappearance. Management should thus be improved in priority and the opportunity to improve drainage could be analysed.

Perennial weeds can sometimes create problems in both permanent and temporary grasslands. Prevention measures are priority. Since most species other than couch grass can produce a lot of viable seeds per plant each year, management should prevent seed production and spreading. One way to do this is by alternating grazing and cutting, by cutting refuse in grazed swards, and by adopting a fast cutting regime, such as taking four cuts per year in exclusively cut grasslands.

Animals can be used as 'collaborators' of the grazing system manager to control weeds. Examples, grazed in rotational stocking system, are given below. In permanent grasslands, sheep and especially goats can help control docks. Donkeys can be used to control thistle as they can browse thistle stems and flowers if they have no other choice.

In arable lands, when temporary grasslands are part of an annual crop rotation, the period between crop harvest and grassland sowing can be used for controlling docks, thistle and couch grass by mechanical means. Techniques differ according to species. Dock roots should never be cut into small pieces by the superficial action of a rotary harrow or rotavator because each fragment could produce a new plant. This type of action would actually multiply dock plants, and even sometimes exponentially. The ideal technique consists in using a harrow equipped with wide wing coulters that cut roots at last seven cm deep. The deep tap root of dock cannot produce new stems when cut at this depth. The upper part of the roots could sprout again. They should be dragged away on the soil surface by tine harrow and dried out. In case of a heavy invasion, these roots can even be collected on the surface. Couch grass rhizomes can also be dragged away on the soil surface by tine harrow and dried out. Thistle cannot be controlled in a similar manner because its rhizome can be as deep as two meters. There is, however, a very efficient method for eliminating thistle entirely. A lucerne/grass mixture cut four times a year for two to three years will totally eliminate creeping thistle. This weed species does not tolerate frequent cuts, and its regrowth after a cut is slow, and after some cuts it is totally dominated by the fast-growing forage mixture. This technique is also efficient on couch grass.

Herbicide use should be used as a final resort if a farmer decides to apply these products. Herbicides can be effective on dock, stinging nettle, ragwort and couch. They are frequently not effective on thistle because after foliage destruction, the strong rhizomes produce new stems again the next year.

For more information: MeadowMania, information on common grassland weeds: www.meadowmania.co.uk/news/information-on-the-common-grassland-weeds/ (Technical and scientific syntheses and scientific papers are available on this site.)

Chapter 2

Grazing management

Fergus Bogue, Michael O'Donovan, Leanne Bastiaansen-Aantjes and Agnes van den Pol-van Dasselaar

When it comes to maximising the benefits of grazing, there is a lot to learn from Irish grazing management. Therefore, this chapter is chiefly based on guidelines and insights from Ireland, where the majority of herds are spring-calving. Even though grass-based and grassland-based systems are in general seen as low cost systems, there is a tendency in Europe for ruminant production systems to intensify, leading to more concentrates and maize in cows' rations along with less grass in the rations and less grazing. Furthermore, the assumed economic benefits of grassland-based systems are not achievable in practice in some European areas due to farm and pedoclimatic conditions or are perceived as impossible by farmers. They choose to be less grassland-based and transform part of their grasslands to more profitable systems. However, the principles presented in this chapter are useful knowledge for all farmers that practise grazing, either full grazing or restricted grazing with supplementary feeding.

Chapters 2.1, 2.3, 2.4 and 2.5 are based on *Grazing Guide 2 – Joint publication between Teagasc and Irish Farmers Journal* (2016), 90 pages, www.farmersjournal.ie

▸▸ Grass supply

The content of grass

The protein content of purchased rations and, increasingly, the UFL (energy value) of purchased rations can be rolled off the tongue of any farmer in the country. But knowledge about the nutritional attributes of grazed grass, which makes up 60% to 80% of the total dry matter intake of most ruminants, is less well known.

Grass can be divided into its water and dry matter content. As you can see below in Figure 2.1, 100 kg of grass will contain approximately 83 kg of water. But it is the dry matter that contains the key nutrients that the animal needs. The dry matter can be divided into cell wall and cell contents. The cell wall of grass is the fibre content, while the cell contents include sugar, protein, fats, minerals and other compounds.

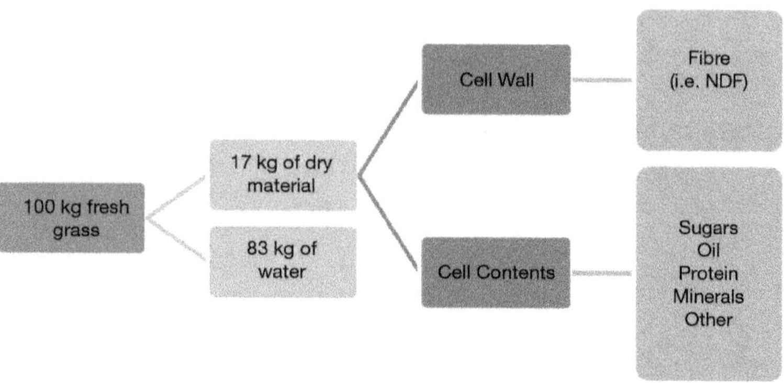

Figure 2.1. The contents of grass.

Energy

The energy in grass comes primarily from the sugar and fibre content, with some energy from oil and protein. The higher the proportion of leaf in grass (ideally over 80%), the higher the energy content coming from sugars and digestible fibre. Fibre is a key supplier of the energy in grass but it needs to be quality fibre. As the proportion of stem in the grass plant increases, the digestibility of the fibre decreases and consequently the energy content decreases. Therefore, grazing leafy grass is ideal for maximising performance.

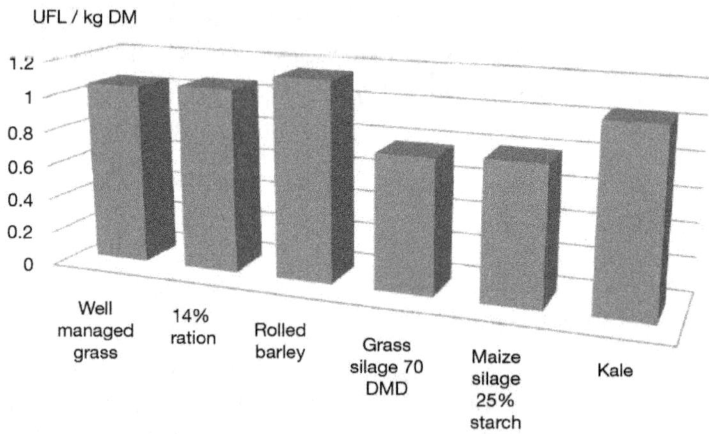

Figure 2.2. The energy content of various feeds.

The energy content of grazed grass varies from 1.05 UFL/kg DM for leafy fresh spring grass to 0.85 UFL/kg DM for very stemmy grass in the autumn. This compares well to other feeds (Figure 2.2). Grass energy content is controlled by maintaining swards with high quality grass content but equally important is good grassland management – grazing out of paddocks in springtime, maintaining a 21-day rotation through the main grazing season, avoiding grazing heavy covers of grass and grazing to 4 cm.

Dairy cow energy demands can be met by a grass-only diet throughout the main grazing season, with some supplementation needed at the shoulders of the year when grass supply is limited. Similarly, the energy demands of suckler cows, calves, yearlings and finishing steers and heifers can be met by a grass-only diet. Dairy production can increase through supplementation, but these come with a cost.

Protein

The protein in any feed can be divided into the quantity and quality of the protein. The quantity of protein in grass typically varies from 16-28%, depending on the sward type and its botanical composition, growth stage, fertiliser regime and time of the year. Occasionally, protein levels in grass dip as low as 11% to 12%. This can happen during a period of stress on the grass plant (e.g. a drought). Quality of protein is defined by systems that account for the quantity of protein that can be utilised by the animal (not all protein in a feedstuff is utilisable by the animal).

How much protein does the animal need? Protein is a key nutrient for appetite, milk production, reproduction and growth. Young, growing cattle and lactating cows need the most protein. Young stock needs 13% to 15% crude protein (CP) in the diet, lactating cows with full grazing need 14% to 17% CP depending on yield, and finishing cattle need 11% to 12% CP. Based on this information, it is clear that the quantity of protein in grass is in excess of requirements in the case of full grazing. In fact, there is an energy cost to the animal excreting the excess protein in grass. Therefore, avoid feeding supplementary protein on grass. There is a cost in buying it, a cost in excreting the excess protein and a large environmental cost (nitrates in water, nitrous oxide as GHG).

Protein quality tends not to be an issue for young stock, suckler cows or finishing cattle on grass. But for freshly calved cows in springtime, there is a need for some quality protein from ration for the first six weeks of lactation. And while autumn grass has adequate protein for late-lactation spring calving cows, freshly calved autumn calving cows need some quality protein in the ration to meet their requirements.

Fibre

The rumen is the engine house of the ruminant animal, so maintaining a healthy rumen is key to good performance. Ruminants are unique in their ability to digest fibre from grass and other forages, and fibre is important in maintaining a healthy rumen. Cows have a specific requirement for fibre. When this requirement is not met, rumen pH becomes unstable and animal performance suffers.

Too little fibre is a problem, but so is too much fibre. Too much fibre reduces dry matter intake, energy intake, and body condition gain, leading to production losses. Grass fibre content is defined by the neutral detergent fibre content (NDF, %). The NDF content of grazed grass varies from 35% for leafy fresh spring grass to 50% for stemmy grass. Dairy cows need a minimum of 30% fibre (NDF) to maintain a healthy rumen. Beef cattle can thrive with much lower levels of fibre in the diet. There is generally more than adequate fibre in grazed grass. Rumen pH (level of acidity) tends to be lower in grazing diets but research work from Northern Ireland, New Zealand and Australia indicates that feeding additional roughage has no impact on animal performance.

Minerals

The mineral content of grazed grass can be divided into major elements (including calcium, phosphorus, magnesium, sodium and sulphur) and trace elements (including copper, selenium, iodine, cobalt, manganese and zinc). As is evident from Figure 2.3, major elements tend to be well supplied in grazed grass but deficiencies of major elements do occur, such as magnesium during the tetany period or phosphorus on deficient soils.

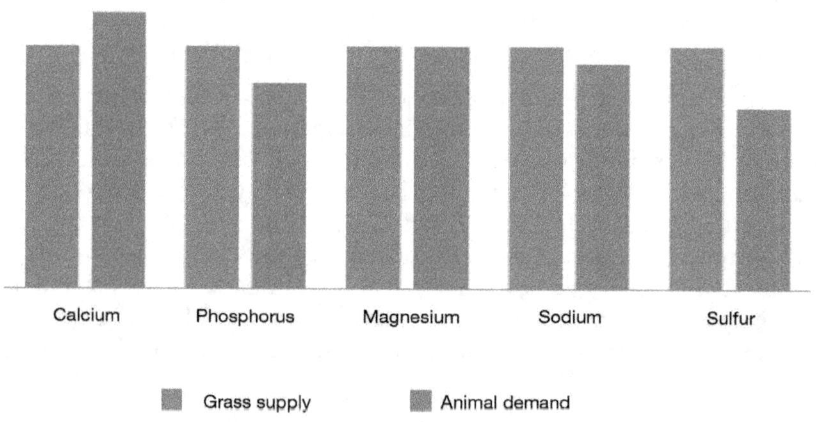

Figure 2.3. Grass supply relative to dairy cow demand for major elements.

Trace element levels in grass are low (Figure 2.4) and consequently need to be supplemented at key periods during the year, including pre-calving, post-calving, during the tetany period and during the breeding season.

It can be useful to have grass analysed every three to four years to take the guess work out of mineral supplementation. This will establish the mineral status of the grass and the presence of any antagonists such as molybdenum and iron. While grass is deficient in trace elements, over-supplementing with trace elements can cause more problems than it will solve (i.e. toxicity).

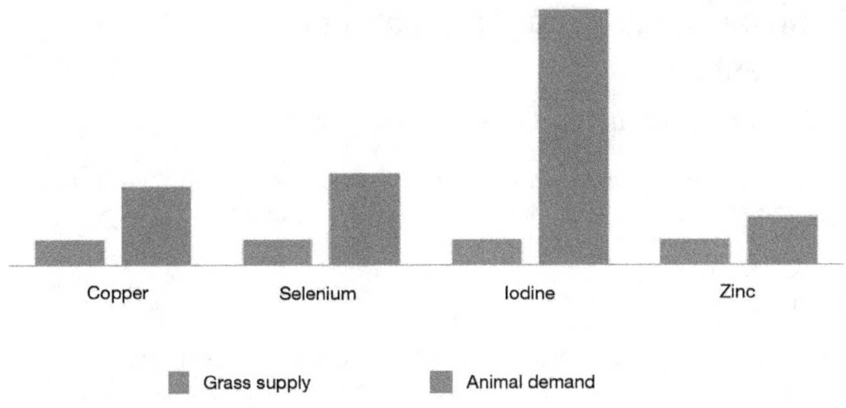

Figure 2.4. Grass supply relative to dairy cow demand for trace elements.

Cost

Grazed grass remains the cheapest feedstuff to produce. Examples from Ireland show that grass costs €80/1,000 units of energy (Figure 2.5). It is 2.5 times cheaper than grass silage and 3.5 times cheaper than a ration at €295/tonne.

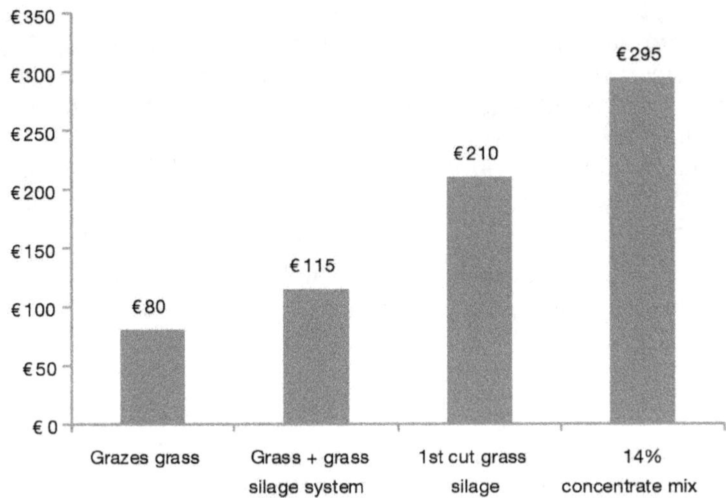

Figure 2.5. The cost of grazed grass relative to grass silage and concentrate feeds (€/1,000 UFL) in Ireland.

▶▶ Main parasitic diseases encountered on the pasture

P. Gauder, D. Jacquet and L. Bastiaansen-Aantjes

Grazing cattle are subject to various parasitic diseases. Most often it is possible to work on the grassland management to reduce infestations.

Distomatosis

This is a subclinical disease that affects all categories of animals and can have serious economic consequences. It is caused by the liver fluke (*Fasciola hepatica*). The parasite is ingested by the animal and goes into the liver where it causes lesions. The symptoms appear gradually over several years. They result in a loss of weight and a decrease of milk production and fertility.

Snails and other amphibious molluscs are the intermediate host of liver fluke larvae. In order to reduce the risk of infestation, these intermediates must be reduced on the sward. For this purpose, ensuring a good soil drainage of the meadows is important. Moreover, applying calcium cyanamide may be useful.

Bronchial and gastrointestinal worms

Lung and gastrointestinal worms mainly affect young animals. Lungworms (*Dictylocaulus viviparus*) cause bronchial lesions. Symptoms are cough and breathing difficulties. The gastrointestinal worms of ruminants are located in the abomasum (*Haemonchus contortus, Oestertagia*) and the small intestine (*Cooperia, Trichostrongylus Axei, Nematodirus, Strongilidés, Monieza, Oesophagostomum*). They cause inflammation of the mucous membranes, nutrient malabsorption, diarrhoea and lower performance.

When turning cattle out to pasture, a bolus in the rumen or ivermectin treatments are often used to prevent these parasites.

Coccidiosis

Coccidiosis is caused by a protozoan (*Eimeria bovis, zurnii*). This parasitic contamination happens mostly in summer and especially affects young animals that get infested by ingesting the parasite eggs. The presence of wet areas in the meadow favours the development of this parasite. Nevertheless, the pasture is not the only place where cattle catch the parasite – it can also happen in the barn. In the animal, the parasite is found in the colon, the caecum and the rumen where it causes destructive lesions of mucous membranes. The animals suffer from haemorrhagic diarrhoea, weight loss and anaemia.

To prevent it, the formation of wet spots on the meadow should be avoided. If possible, water points should be moved to disrupt the development of this parasite in these areas. The most-used treatments are injections of sulphonamides or ingestion of amprolium through food supplementation.

Other pathologies encountered by young calves

Young calves can suffer from enterotoxaemia, which is caused by the absorption of large volumes of toxins produced by *Clostridum perfringens* from the intestines. There are few symptoms and death can occur within a few days. Unfortunately, there is no effective treatment against it. Medical prophylaxis (vaccination) and food prophylaxis remain the only cures. With regard to food, it is important to provide enough fibrous food in the ration of the suckler cows and avoid any abrupt dietary changes. Supplementation to keep a balanced ration may also help.

Calves can also have diarrhoea as soon as they are put on the meadow. This can show up as white faeces, which is a symptom of poor milk digestion. It happens when cows consume young grass with a low dry matter content and without structure. Their milk then contains too many long chain fatty acids, which causes a disruption of their digestion.

Diarrhoea in calves can also result in black faeces when the cows consume a very high-quality grass and produce milk that is rich in fat and protein that will cause indigestion in calves.

▸▸ Main grazing parameters

F. Bogue, M. O'Donovan, L. Bastiaansen-Aantjes and A. van den Pol-van Dasselaar

Grass height and stocking rate influence grass intake and livestock production. To quantify the effect of grass height and stocking rate, a few parameters need to be known.

LSU/ha

LSU/ha, also known as stocking rate, is the total of all livestock units at grass divided by the measured area in hectares. Each livestock type has its own livestock units:
– Dairy milking cows, suckler cows, stock bulls, dry cows: 1.0
– 0-6 months old: 0.1
– 6-12 months old: 0.3
– 1 -2 years old: 0.7
– >2 years old: 1.0
– Lactating ewes: 0.25
– Dry ewes/hoggets/goats: 0.15
– Lambs: 0.0
– Horses: 1.0

Cover/LSU

Cover/LSU is Average Farm Cover divided by the stocking rate on the farm. Cover/LSU is used to compare farms with varying stocking rates. Sometimes this is referred as Cover/Cow, where a cow is one livestock unit. In Ireland, this parameter is primarily used in the summer between April and August.

Average farm cover

Average farm cover (AFC) is the average amount of grass on each hectare of the farm and is expressed in kilograms of dry matter per hectare (kg DM/ha). To calculate the average farm cover, multiply the paddock area by the cover in each paddock, for all paddocks measured during a walk, and then divide by the total ha of all paddocks measured.

Average farm cover is an important parameter in the spring and autumn. Typically, AFC is high in early spring and reduces towards the end of the first rotation in April. Between April and August, AFC needs to be maintained. From late August to September, target AFC is increased so that this extra grass can be grazed in October and November when demand is higher than growth.

Rotation length

The rotation length is the number of days it takes the herd to rotationally graze the whole farm. Rotation length can be calculated by dividing the total area available for grazing by the number of hectares grazed per day.

Why measure grass

The potential to achieve high levels of productivity from grazed grass gives farmers a major competitive advantage over global counterparts. Existing research clearly shows that farms that grow more grass have lower costs and higher profits. On average, the cost of producing 1 kg of live weight gain or 1 kg of milk solids from grazed grass is 80% to 85% less when compared with an intensive concentrate-based system (grassland database).

Decision support tools like PastureBase Ireland aim to help farmers ensure that they are exploiting the full potential of grazed grass on their farms, irrespective of production system or land type. Land type or location is often seen as a barrier to adopting good grassland management practices. While Irish farmers use some purchased feed, the majority of weight gain or milk is produced from mainly grazed grass.

There are still a number of simple steps that farmers can take to improve grass growth, grass quality and grass utilisation. Getting livestock out to grass early and ensuring an adequate supply of good-quality leafy grass is available throughout the grazing season

is key to obtaining high levels of animal performance. Measuring grass is important to achieve this. The main benefits from measuring grass are to:
– Maximise the proportion of grazed grass in the diet.
– Maximise pasture re-growth rates.
– Improve pasture quality and feed more grass at a higher quality.
– Graze more grass in the spring and autumn, shorten the winter period.
– Achieve target average farm covers at key times during the year.

Table 2.1. Guidelines for cover/LSU, AFC and rotation length in Ireland.

Guidelines for Cover/LU, AFC and Rotation Length

Date	Cover/LU Kg DM	Average Farm Cover (AFC) Kg DM/Ha	Rotation Length Days
STOCKING RATE OF 2.5 LU/HA			
1st February	360	900	100
1st March	280	700	50
4th April	200	500	18-26
May, June, July	170	425	18-24
Mid - August	200	500	25
1st September	300	750	30
Mid-September	400-450	1,000-1,100	35
1st October	400	1,000	40
1st November	60% of your grazing platform should be closed for Spring at this stage		
Fully Housed		550-600	
STOCKING RATE OF 3.0 LU/HA			
1st February	330	1000	100
1st March	250	750	50
4th April	185	550	18-26
May, June, July	170	510	18-24
Mid - August	250	750	25
1st September	330	990	30
Mid-September	370	1100	35
1st October	380	1150	40
1st November	65% of your grazing platform should be closed for Spring at this stage		
Fully Housed		600-650	
STOCKING RATE OF 3.5 LU/HA			
1st February	285	1000	100
1st March	230	800	50
4th April	185	650	18-26
May, June, July	170	600	18-24
Mid - August	220	770	25
1st September	280	980	30
Mid-September	340	1200	35
1st October	335	1175	40
1st November	70% of your grazing platform should be closed for Spring at this stage		
Fully Housed		700-750	

How to grow more grass

Grazed grass is, and will continue to be, the cheapest animal feed for meat and milk production. The land's ability to produce excellent quality grass is the primary competitive advantage over other dairy farmers. To optimise profitability, producers must maximise the proportion of grazed grass in animal's diet.

The key performance indicators in relation to grassland management are as follows:
– A long grazing season will maximise your profitability and competitiveness. Extend the grazing season in early spring and late autumn. Grass budgeting is an

essential tool in achieving a long grazing season. Paddocks should be grazed to low post grazing heights (e.g. 3.5 cm) in early spring to condition swards for subsequent grazing rotations. On/off grazing is a strategy to increase the proportion of grazed grass in the animal's diet during periods of wet weather.

– In Ireland, it is shown that increasing the proportion of grazed grass in the diet of a dairy cow by 10% reduces costs of production by 2.5 cent/litre (2011); every extra tonne of grass DM is worth €173/ha to a dairy farm and €105/ha on a drystock farm (2015); increasing the grazing length by 30 days will reduce production costs by 1 cent/litre (2015). To achieve this, ensure your cows' calving pattern is matched to the start of the grass growing season. Begin calving at the onset of grass growth. Typically this should result in most calves being born between 10 February and 1 March (six weeks before growth meets demand). Use a decision support tool such as the spring rotation planner on PastureBase Ireland and stick to daily area allocations as planned, graze 30% in February, 66% by 17 March and target 100% grazed by 6 April (adjust these dates for later turnout regions).

– Match your stocking rate to the growth potential of your swards.

– Ensure perennial ryegrass dominates all swards.

– Target farm DM production of 15/16t DM/ha.

– Stock the farm to its grass growth capability, e.g. 5t grass dry matter consumed per cow, grass yield 14t/ha = 2.8 cows/ha.

– Maximise the productivity of your swards by improving soil fertility.

– Soil sample one third of the farm each year. If there has been no sampling for several years, consider getting the whole farm sampled.

– Apply P, K and lime as recommended.

– Maximise the productivity of your swards through timely reseeding.

– Reseed in spring if possible.

– Target a 60-day turnaround time from seeding to first grazing.

– Ensure that recommended list varieties are used.

– Use a post-emergence spray at the two-leaf regrowth stage.

– Graze the sward for the first time at 600-700 kg DM/ha.

– Ensure good grassland management.

– Make use of farm grass cover measurement and grass budgeting.

– Use feed concentrates/high quality silage when short of grass.

– Ensure that farm infrastructure is sufficient to fully utilise the grass grown.

Key performance indicators

It is impossible to manage something that is not measured. Grass covers must be estimated in each individual paddock on the farm, and this information can be used to achieve both short daily and medium term (weekly and monthly) targets that are critical to the success of the system. Such skills can be learned from advisors, through farm discussion groups and through practice and self-training.

▸▸ Management according to the season

*F. Bogue, M. O'Donovan, L. Bastiaansen-Aantjes
and A. van den Pol-van Dasselaar*

Spring

To capitalise on the benefits of grazed grass, dairy cows should be turned out to grass directly after calving, ground conditions permitting. The main objectives of spring grazing management are to:

1. Increase the proportion of grazed grass in the diet of the dairy cow

2. Condition swards for subsequent grazing rotations.

Farm cover at turnout should be approximately 800 to 900 kg DM/ha, depending on the mean calving date – an earlier calving date equates to higher animal demand and the need for a higher opening cover. Aim to offer 1.0 to 1.2 tonne grass DM/cow from turnout until the end of the first rotation; this is achievable on farms where animals are turned out early. Grazed grass and concentrate can be the sole feed with such a system. This allows grass silage to be completely removed from the diet post-calving.

Grass at a reasonable level of utilisation (80%) costs about 7 cents/kg utilisable dry matter (DM) compared with first and second cut grass silage at 15 cents/kg and 18 cents/kg utilisable DM, respectively.

Grazed grass is the highest quality feed on the farm in spring; it is better than silage and equivalent to concentrates. Based on these figures, it is important to increase the grass proportion in the diet of grazing animals.

Spring is the key period to target for several reasons:
– More expensive feeds such as grass silage and concentrates can be replaced by grazed grass.
– Early spring grazing increases grass quality in the second, third and subsequent grazing rotations.
– Grazing more grass in spring will increase the total quantity of grass grown on the farm on an annual basis.

During the early grazing season (February or March), a balance must be struck between feeding the animal adequately to maintain high animal performance and conditioning the sward for the late spring/summer grazing season. In the first rotation, the key is to graze paddocks out to 3.5 to 4 cm and set up paddocks for following rotations. By doing this, grass digestibility can be increased by four to six units in May and June.

The target is to achieve a 280 to 300-day grazing season. Across Ireland, the average grazing season length of PastureBase Ireland farmers is 240 days.

Animal performance from early turnout increases substantially for both finishing and store cattle as well as for dairy cows.

Obstacles

One of the main obstacles to achieving more days at grass in early spring is poor soil conditions. If animals stay in the paddock, treading damage on soils can result in reduced growth rates during subsequent grazing rotations. Even on dry soil paddocks which are severely damaged by poaching, grass supply can be reduced by 30% at the next grazing. It takes even longer for heavy soil paddocks to recover; grass growth can be reduced until the following autumn.

Allowing animals access to pasture for approximately six hours per day –three hours in the morning and three hours in the evening (on/off grazing) – has been shown to maintain high levels of performance when compared with grass silage-based diets and may be a strategy that can be implemented to extend the grazing season length. Animals can adjust their grazing behaviour to ensure that they achieve almost the same intake as they would if grazing on a full-time basis. Once there is sufficient grass on the farm it is important not to offer the animals grass silage when they return indoors as this will decrease their appetite, reduce their grazing efficiency and compromise grass utilisation.

Things to do

– Maximise early spring grazing in the diet of freshly calved dairy and suckler cows or priority cattle (e.g. replacement heifers)
– Graze paddocks to 3.5 cm in the first rotation
– Implement on/off grazing or remove stock from grass to prevent damage
– Achieve spring rotation planner targets (e.g. 30% by 1 March, 60% by 17 March, 100% by 4 April.

Getting the turnout date right

The aim in spring is to increase the proportion of grass in the diet of the grazing animal while at the same time budgeting so that there is enough grass until the start of the second grazing rotation in early to mid-April. Spring grazing should start in February and continue until early to mid-April. This varies from farm to farm, but the most important aspect of grazing management is to make good use of spring grass. It is also worth noting that the end of the first rotation can vary from year to year and is dependent on grass growth.

Get priority stock out

All animals in the herd do not have to be turned out at the same time. Groups of animals should be prioritised for early turnout, i.e. those that will benefit most from high quality spring grass such as lactating cows, replacement heifers or steers.

Magic day

This is the day your farm is growing as much grass as you need (i.e. grass growth equals herd demand). This is dependent on the farm stocking rate and the level of supplement used. For some farmers it will be the end of March while for others it will be the middle of April. You will only be able to establish this date after a number of years of continuous grass measurement.

First rotation

The first grazing rotation should be 40 to 50 days and finish on magic day (e.g. 10 April). This can be extended to 20 April in later growing years or poorer grass growing areas.

Area to graze first

Graze 30% to 40% of the grazing paddocks first to ensure there is enough time to allow re-growth accumulate for the start of the second rotation. Silage ground should be grazed early in the first rotation – this will increase the available grazing area. Also, if grass growth is slow silage paddocks can be regrazed for a second time at the start of the second grazing rotation. This will give the rest of the grazing paddocks a few extra days to grow more grass.

Strip grazing

During the first grazing rotation fresh grass should be allocated daily. For milking cows, fresh grass should be offered after every milking. This will help increase grass utilisation.

Animals should be 'back fenced' – that is, when they are given fresh grass they should not be given access to the area previously grazed. This will reduce poaching damage but it will also allow the grazed areas to start re-growing, thus ensuring grass for the second rotation.

Post-grazing

Post-grazing heights of 3.5 to 4 cm should be targeted during the first grazing rotation. Heights below 3.5 cm will reduce subsequent grass growth while above 4 cm will waste feed and reduce utilisation. It will also reduce grass quality in subsequent rotations.

Late turnout and high grass covers make it difficult to achieve target post-grazing heights. This will often lead to poor grass utilisation and subsequent poor pasture quality in subsequent grazing rotations.

Applying slurry in spring

Slurry is a valuable source of N, P and K and should be applied on the fields/paddocks that need it most and at the time of year that will give you the best response.

All of the P and K in slurry is available to be utilised and fields that are low in both of these nutrients need to be targeted to receive slurry. On a lot of farms, this will be their silage fields as this is where the feed that eventually produces the slurry comes from in the first place.

The time of year that slurry is spread does not affect the availability or utilisation of P and K. This is not the case with N.

Spreading nitrogen fertiliser

Nitrogen fertiliser can provide a boost to spring grass growth, allowing for a greater proportion of the grazing animals' diet to be made up of grass. It can also allow a greater number of animals to be turned out to grass in early spring. Soil temperatures need to be at least 5 °C before there is an adequate response to N and the date at which this occurs can differ from year to year. In general, spreading nitrogen on the farm brings forward the farm's grass growth rate by three weeks.

In good growing conditions, 1 kg of N has the ability to grow 10 to 15 kg of grass DM during February, while in other years there can be little or no grass growth response to the N due to prolonged cold weather into March. Urea is cheaper per kg N than CAN and should be used in spring wherever possible to reduce costs.

Spring rotation planner

The spring rotation planner is a tool that divides the area of your farm into weekly portions and takes the guesswork out of planning the first grazing rotation.

The spring rotation planner will not tell you if you are feeding animals enough grass – you will have to gauge that by walking through your paddocks or fields and assessing either visually or by plate meter measuring if you have enough grass. The spring rotation planner is a simple and effective tool that ensures:
– Sufficient grass is grazed early enough to allow time for re-growth for the second rotation.
– Grass does not run out before the start of the second rotation. A wedge-shaped supply of grass is created, ensuring a continuous supply during the second rotation.

The simple rules are:

Dry farms:
– Turnout early to mid-February
– At least 30% of the farm grazed by 1 March
– 60% of farm grazed by 17 March
– 100% of farm grazed by the end of the first week in April

The date the first rotation ends will vary from farm to farm. Your specific magic day (end of the first rotation) can be established over a number of years by walking your farm and measuring farm covers and grass growth which will help build up a record of what to expect from year to year.

Heavy/slow grass growing farms:
– Turnout late February/early-March
– 30% of the farm grazed by 10 March
– 60% of the farm grazed by 27 March
– 100% of farm grazed by mid-April

In general the dates by which a certain proportion of the farm should be grazed are 10 to 14 days later on heavy farms compared to dry farms.

It is important that these targets are achieved as it will ensure there is sufficient grass available by the start of the second rotation. By grazing a certain amount per week, a wedge shape of grass supply can be achieved. For example, if all animals are let out to grass a couple of days before 30% of the farm should be grazed it means that all these paddocks are re-growing at the same rate. There will be a shortage of grass at the start of the second rotation and then too much grass as all paddocks reach their target pre-grazing yield at the same time.

Week	Total area grazed by week ending (%)	Cumulative area grazed (ha)/week ending	Pregrazing Herbage Mass on April 5th (kg DM/ha)
1st to 7th Feb	7	2.8	
8th to 14th Feb	14	5.6	800 - 1 200
15th to 21st Feb	21	8.4	
22nd to 28th Feb	30	12	
1st to 7th Mar	45	18	
8th to 14th Mar	60	24	400 - 800
15th to 21st Mar	73	29.2	
22nd to 28th Mar	87	34.8	100 - 400
29th Mar to 4th Apr	100	40	

Figure 2.6. Example of a Spring Rotation Planner of a 40 ha dry farm where turnout date is 1 February and the first rotation ends 5 April.

The template of Figure 2.6 can be photocopied and used every spring. For example, if the farm is 50 ha then 50 (ha) multiplied by 7 and divided by 100 (i.e. 7%) is the area which should be grazed by the end of week 1. Using this example, 3.5 ha should be grazed by the end of week 1 – this means 0.5 ha should be grazed every day for the first week (3.5 ha divided by 7 days). Each week can be worked out in the same way. Note: for the first three weeks, the weekly area of the farm which should be grazed is 7%, and from week 4 onwards the area per week increases.

Spring grazing for different animal types

Beef cattle

Research work has shown that animals turned out early to grass in early spring have 6% (+23 kg) higher carcass weight than animals turned out later in spring. This could equate to close to €60 to €70/head.

Turnout of animals should take place during periods of dry weather, with good underfoot conditions to give animals an opportunity to 'settle' and start grazing properly. Early turnout will reduce the accumulation of surpluses during the main grazing season.

Dairy

Each extra day at grass in spring is worth €2.70/cow/day, which is achieved from reduced feed costs and labour input (slurry spreading etc.) and an increase in milk protein concentration.

Dairy cows should be turned out directly post calving. Cows should start grazing lower covers first to get them used to a grass diet again. After a week to ten days they can start grazing heavier covers. Covers over 1600 kg DM/ha (>4 cm) should be grazed by early March at the very latest.

Feeding dairy cows in spring

– Cows reach peak lactation six to eight weeks after calving.
– Peak dry matter intake (DMI) occurs 10 to 12 weeks after calving.
– Cows use their fat reserves to make up the energy deficit in early lactation ('milks off her back').
– Cows calving into a grass-based system have a total DMI of 8 to 11 kg DM/cow/ day during the first week after calving. This increases by 0.75 to 1 kg DM/cow/day up to peak intake which is 16 to18 kg DM (Figure 2.7).
– Care should be taken to ensure that cows do not lose more than half a unit of BCS as cow fertility will suffer if this occurs.
– Winter milk herds can reduce the rate of concentrate supplementation by 1 to 2 kg DM when grass is included in the diet.

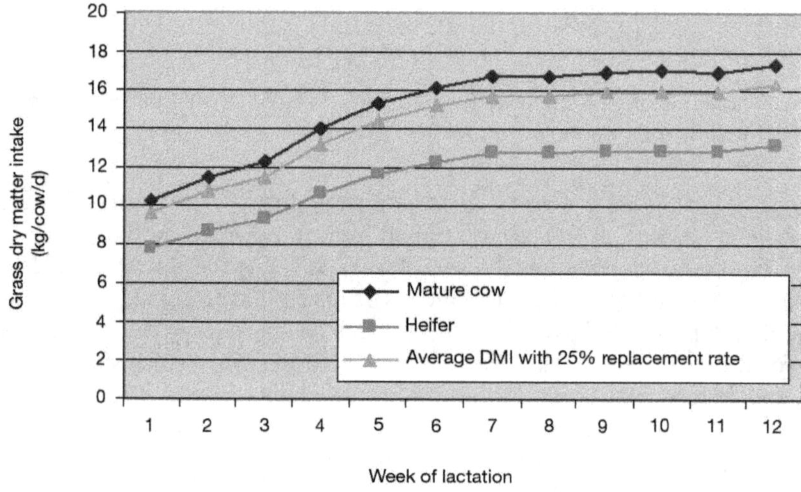

Figure 2.7. Grass dry matter intake in early lactation.

Supplementing dairy cows at grass

– The quantity of grass available on the farm will dictate how much supplementation is offered to the cows.
– It is essential that as well as following the spring rotation planner the farm is walked weekly and a cover completed so that decisions on grass availability, supplementation level etc. can be made.

– Supplementation required = cows energy requirement – grass energy intake.
– In general, the maximum level of supplementation for freshly calved cows in the spring should be 6 kg DM; if the deficit is greater than this, high-quality grass silage should be offered in combination with concentrates.
– As a general rule, six to eight weeks after calving 15 kg DM of grass plus 3 kg DM of concentrate is sufficient for a peak milk solids yield of 1.8 to 2.0 kg/cow/day; concentrate level can be reduced depending on grass supply.
– The protein content of early spring grass is high (>20%), which means that high-protein concentrates are not required when the majority of the diet is made up of grazed grass.
– If supplementing to reduce the risk of grass tetany make sure to check the concentrate composition so that the correct rate is being fed to the covers to offer protection. Cows require 30 g of magnesium or 60 g of calcined magnesite per day. Magnesium can also be supplied through pasture dusting or through treated water.

Sheep and goats

The aim is to have enough grass to match ewes' demand until supply increases and matches demand (magic day). For an early/mid-March lambing flock an opening farm cover of 600 to 700 kg grass DM/ha or 20 to 25 days ahead is recommended. Why?
– 10 ewes/ha with average demand of 2.5 kg DM/head/day in early lactation
– 10 x 2.5 = daily requirement of 25 kg DM/day
– 650 ÷ 25 = 26 days ahead

If we estimate average grass growth rate of 15 kg DM/ha/day in early/mid-March this will add another 10 to 15 days, so this means there are 35 to 40 days which should extend to mid-April (magic day).

Feeding of sheep and goats is often based on information derived from cattle. But they are different in many metabolic indicators (see Table 2.2).

Table 2.2. Metabolic indicators of cattle, sheep and goats.

	Cattle	Sheep	Goats
Live weight (kg)	600	45	45
Metabolic weight (MW) (MW = $W^{0.75}$) (kg)	121.2	17.4	17.4
Net Metabolic Energy (NME) (Kcal/kg $P^{0.75}$)	70	65	56
Total Net Metabolic Energy (tNME) (Kcal)	8484	1131	975
Volume of the digestive system* (VDS) (ml)	90000	6750	6750
Digestive capacity (DC)(= VDS/tNME)	10.6	6.0	7.7

*VDS equals 13% to 18% of total body volume.

The use of pastures by sheep and goats depends on several factors, particularly their dry matter intake and the quality and quantity of pastures.

Low-speed ruminal degradation feeds (cellulose, hemicellulose, insoluble proteins and starches) having a longer time of food retention in the rumen and are better

utilised by cattle than by goats and sheep. This because cows have a larger rumen than sheep and goats. Sheep and goats balance these disadvantages by:
– a higher voluntary intake of feed (5% to 7% of their weight vs 3% to 4% in cattle). Other things being equal, sheep and goats show a DM intake of 5 to 5.5 kg DM per 100 kg live weight vs 2.8 to 3.2 in cattle during lactation.
– selecting species with a higher rate of rumen degradation (less fibrous plants, leaves, shoots) in pastures. This option is facilitated by the particular conformation of the muzzle and the greater lingual and labial mobility than cattle.
– Sheep and goats take much longer than cattle to ingest and ruminate (about 10 times more). In fact, they have a less powerful chewing activity, and they need to grind foods finer to allow transit in the digestive system. Consequently, intake is more influenced by fibre dimension in sheep and goats than in cattle.

DM intake depends on the metabolic weight of the sheep or goat: during full lactation, it varies from 2.5 kg DM in a sheep of 50 kg (equal to 5 kg/100 kg) to 2.7 kg DM in a sheep of 60 kg (equal to 5.4 kg DM/100 kg). Dairy sheep usually have higher DM intakes than meat sheep. In dairy sheep, voluntary intake changes in relation to the physiological stage: in a 50 kg sheep, voluntary intake changes from 1.6 (in the last part of pregnancy) to 2.4 kg DM (at the second to third month of lactation). At the same lactation stage, DM intake depends on milk production. During pregnancy, DM intake depends on the number of lambs and becomes lower with the increasing number of lambs. Adverse environmental factors (wind, sun, rain, cold temperatures) during grazing can influence voluntary intake. Voluntary intake in sheep and goats is also influenced by the pasture quality (Table 2.3), and especially its digestibility: the higher the digestibility, the higher the intake.

Table 2.3. Effect of different grass characteristics on voluntary intake.

Feed	Voluntary DM intake (kg 100 kg live weight)
High quality pastures	5.0
Medium quality pastures, excellent maize silage, high quality haylage, high quality hayw	4.0-5.0
Poor pastures and green forage, medium quality maize silage/haylage/hay	2.5-4.0
Very poor pastures and green forage, poor hay/silage, straw	1.7-2.5

Note that sheep and goats often are bred in hilly and mountainous areas and have high energy requirements for walking in addition to basic energy requirements (0.53 UFL for a 40 kg sheep, and +0.09 UFL for every +10 kg). On average, they need 25% more energy for grazing in good pastures and higher requirements in poor and non-homogeneous pastures.

The optimal height of pastures for sheep is about 6 cm in spring with continuous grazing and 7 cm to 8 cm with rotational grazing to keep herbage green, leafy and with a good nutritional value.

Issues

– Insufficient area closed in autumn to build covers for spring.
– No N applied to boost covers and enhance March growth rates.

Summer

Managing grass supply: balancing grass supply with grass demand

The key to mid-season or summer grazing is to ensure a constant supply of high-quality grass ahead of the herd. High performance can be achieved from a grass-only diet once the correct pre-grazing yield is offered. Allowing pre-grazing yield to exceed recommended levels leads to a decline in grass quality, resulting in poor milk solids yield or poor rates of weight gain.

Finishing the first rotation on time is critical for mid-season grass supply and quality. It will ensure that grass will be more easily managed in the second and subsequent grazing rotations. Finishing the first rotation too early will mean animals are grazing on too short a rotation from April onwards and grass will be in short supply. Finishing the first rotation too late means there will be surplus grass, post-grazing residuals will be difficult to achieve and quality will be affected for the rest of the year. The biggest problem on farms is that the first rotation starts and ends too late.

Avoid wasting offered grass to the herd

Too often excessive grass is offered to the grazing herd. High grass utilisation (>80%) is possible when pre-grazing yields are at levels that the grazing animals can graze out well.

The key focus during the main grazing season is to offer high-quality/leafy material to the grazing herds as often as possible. Grazing animals respond positively to high-quality grass and it is far easier for them to graze swards with a range of 1,400 to 1,600 kg DM/ha (8 to 10 cm) than swards of 2,000 to 2,500 kg DM/ha (12+ cm).

General points

– Grazing to 3.5 to 4 cm in the first rotation provides a platform for excellent quality re-growth in second and subsequent rotations.
– The ideal pre-grazing yield for maximum animal performance in summer is 1,400 to 1,600 kg DM/ha (8 to 10 cm).
– Under-grazing leads to a greater proportion of stem. This will lower grass quality and animal performance.
– Avoid turning stock into overly heavy covers. React quickly to surplus grass and harvest as baled silage.

Grass quality

Grass quality is indicated by grass digestibility. Maintaining high-quality pastures with a high proportion of leaf is critically important during the summer/main grazing period.

– Grass quality influences dry matter intake and production performance.
– High digestibility grass has a high intake potential and high energy content.
– High digestibility grass is characterised by:
– high leaf content and low true stem content
– high protein concentration and medium fibre concentration
– short-medium regrowth interval and low-medium pre-grazing herbage mass

High production performance from grazing livestock has a major influence on farm profitability. This is achieved by ensuring high intakes of high quality grass. When correctly managed, grazed grass is of high nutritive value. Grass quality, as indicated by organic matter digestibility (OMD), can be maintained at a high level throughout the grazing season under good management practices.

Digestibility is the biggest factor determining the energy content of grass. The digestibility of grass exhibits a characteristic pattern of change during the year. The highest values are obtained in spring (80% to 85%), when the grass plant is leafy; they are lowest in mid-summer (78% to 80%) when the grass plant is in a reproductive (flowering) phase; they are intermediate in autumn (79% to 81%).

When the quantity of grass is not limited, the primary factor influencing intake by grazing animals relates to the digestibility of the available grass. Changes in digestibility are associated with changes in the amount of green leaf, mature stem and dead material. As the grass plant matures, the proportion of leaf to stem decreases; this is associated with a decrease in the ratio of cell contents to cell wall constituents, especially in the stem fraction.

Green leaf is highly digestible. Animals will select green leaf over stem. Grass will be low in digestibility when it has a lot of stem, flower heads and dead material. The problem with this type of grass is that the intake will be low because animals do not like what is offered. They find it difficult to graze and they do not digest it well, which means they extract fewer nutrients from it.

Visual assessment

A practical measurement of grass quality is a quick assessment of the grass being offered to animals. A small representative sample of grass can be cut at the predicted post-grazing height. This can then be visually assessed for sward morphology. The following questions should be considered.

What colour is the sample? Bright green swards (more leaf) are an indicator of high digestibility, while pale green/yellow swards (more stem and dead material) are less digestible.

What is the texture of the sample? Green leaf has a soft texture which is easier for animals to consume and digest, while stem has a coarse texture which is more difficult for animals to consume and digest.

What is the leaf to stem ratio? A manual separation of the sward components can be used to give an indication of grass quality. Well-managed grazed swards will contain between 70% to 80% green leaf, 15% to 20% green pseudostem and less than 5% mature stem and 5% dead material. This is important as every 5% increase in sward leaf content results in a 1% increase in digestibility.

Leaf stage/pre-grazing herbage mass

Well-managed grazing swards which are grazed between 1400 to 1600 kg DM/ha pre-grazing herbage mass (PGHM) are generally at the 2.5 to 3 leaf stage of growth. This means that the third leaf on a tiller is emerging or fully emerged. Grazing at this stage of growth is optimal as it maximises green leaf content in the sward and avoids leaf senescence, which would occur if the sward is left to grow. Grass quality can be maintained by grazing swards with a PGHM to a post-grazing sward height of 4 cm throughout the summer.

Managing grass supply during the main grazing season

The objective during the main grazing season (May to August) is to achieve high performance from a grass-only diet. High animal performance will be achieved by maintaining a consistent grass supply for the herd and monitoring farm grass cover weekly. This monitoring will be the basis for a grazing wedge. This grazing wedge will allow decisions to be made to alter grass supply early; for example, adjusting stocking rates or removing surplus grass.

The grazing wedge: understanding your wedge

The line on your wedge graph is drawn from the ideal pre-grazing yield (1400 to 1600 kg DM/ha) to the target post-grazing residual (100 kg DM/ha). A perfect wedge, such as the one below, is where all the paddocks meet the line; in other words, all paddocks are on target, where there are no surpluses (paddocks above the line) and no deficits (paddocks below the line).

Frequently, this is not the case. The following sequence of graphs gives various scenarios that may arise and outlines actions that should be taken to correct surpluses or deficits.

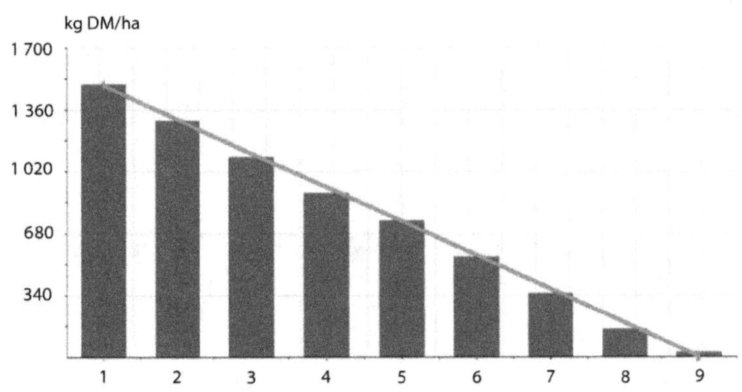

Surplus wedges

From this wedge it is evident that this farmer has a surplus on the farm: the first four paddocks are over the demand line. Pre-grazing yield is well above the 1400 to1600 kg DM/ha target. Residuals (post-grazing yields) are being achieved in this example, but often when grazing, high pre-grazing yield paddocks are not grazed out properly and residuals are too high, which can affect subsequent sward quality.

The following should be kept in mind when dealing with surplus grass:
— Remove surplus paddocks as silage. This should be completed as soon as possible (or once the paddock reaches 2,500 kg DM/ha) so that the paddocks will be back in the grazing rotation as quickly as possible.
— Do not put off addressing high grass growth.
— If the grass in the paddock is not too 'strong' get other animals to graze it, such as heifers/dry cows.

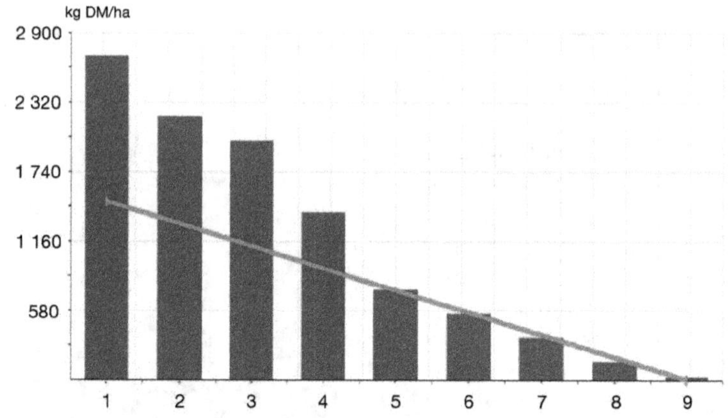

— Do not increase the stocking rate too much on the grazing area by closing too many paddocks for long-term silage.
— Caution should be exercised so that excessive grass is not removed resulting in a deficit.
— Removing surplus grass as soon as it is identified will result in the area being included in the grazing round and therefore making it available to cope with a slowing of pasture growth.

Deficit wedges

The next graph is an example of a deficit where there is not enough grass available to meet target pre-grazing yields.

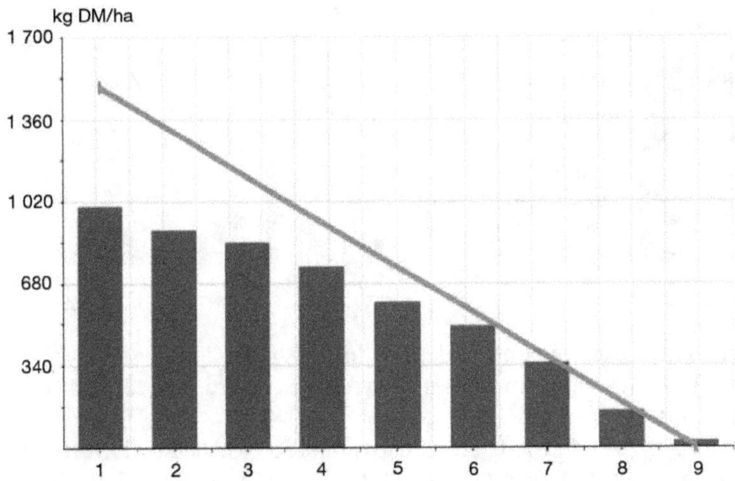

It is clear there is a large deficit on this farm as the first six paddocks on the graph are below their targets. The farmer is currently grazing pre-grazing yields of 1,000 kg DM/ha, well below the 1400 to 1600 kg DM/ha pre-grazing yield target. Extreme action will need to be taken to rectify this problem. Weekly measurement and acting on the information recorded could have prevented this situation arising as the farmer would have been able to see this deficit occurring before it actually happened.

The following should be kept in mind when dealing with deficits:
— In all cases before 'magic day' (the day when grass supply equals grass demand) do not speed up the round.
— After 'magic day' consider increasing the grazing area/day during the deficit period if soil temperatures have continued to rise and pasture growth is increasing.
— Supplement with concentrates or grass silage (preferably high-quality baled silage that was previously removed as surplus as it will be of better quality than pit silage).
— Re-graze the area closed for silage once the pre-grazing yield is not excessive. A strip wire should be used in this situation.

Difficult to interpret wedges

This farmer has a surplus in the first two paddocks, followed by three paddocks which are below target, while the next three are above target. In this situation the farmer should graze the first paddock provided the grass is of high quality. This will give the paddocks below the demand line more time to grow and achieve their target pre-grazing yield. However, if they are below target at grazing, the following paddocks should have sufficient grass to ensure a deficit does not occur on the farm. The farmer should also ensure that the post-grazing height is being maintained in each paddock. This situation is best dealt with by walking the farm in a few days' time to see how much grass is growing and if supplementation needs to be introduced or not.

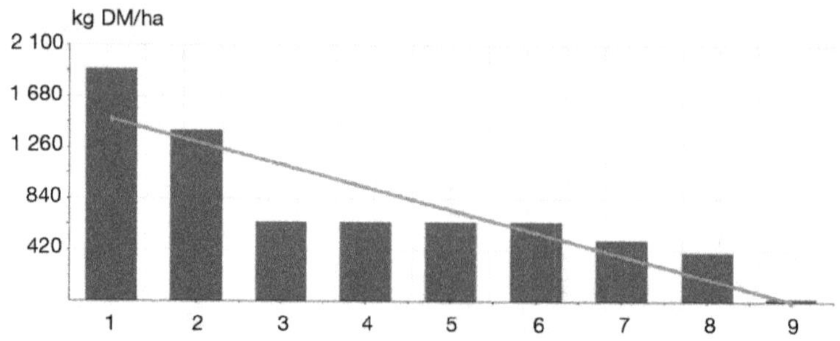

The key grassland management guidelines for this period are:
– Aim to consistently offer animals a sward where there is green leaf down to the base and very little stem.
– Ideal pre-grazing yield targets are 8 to 10 cm (1300 to 1600 kg DM/ha). Maximum pre-grazing yield is 10 to 11 cm (1600 to 1800 kg DM/ha). If this is exceeded, excess grass should be removed as baled silage (depending on grass supply).
– Rotation length should be maintained at 18 to 21 days.
– Early identification of surpluses by measuring grass weekly may reduce or eliminate the requirement for topping.
– Use the grass wedge to identify grass surpluses and deficits.
– These surpluses can be removed as round bale silage, which will replace the necessity to top pastures. They will also act as a source of extra feed if winter is longer than expected.
– Average farm cover per cow needs to be maintained at 160 to 190 kg/DM/day.
– If pastures have high post-grazing residuals (400 to 500 kg DM/ha) or high post grazing height (>6 cm), they should be topped. Pasture topping should take place early (mid-May) rather than late in the season.

Sheep

The key guidelines to maintain quality in pasture grazed by sheep during the mid-season or summer months are:
– Quality versus quantity
– Pre-grazing height = 7 to 9 cm (1000 to 1500 kg DM/ha)
– Leaf = digestibility
– Post-grazing height
– Pre-weaning: 4.0-4.5 cm
– Post-weaning: 5.0-5.5 cm
– Ewes: 4.0-4.5 cm

Autumn

There is a lot of potential to make better use of grass on farms in autumn. Every extra tonne of grass utilised is worth €173/ha. Utilising extra grass and lengthening the grazing season should be your key objective in autumn. The focus of autumn grazing management is to increase the number of days at grass and animal performance, as well as to set the farm up during the final rotation to grow grass over winter and provide grass the following spring.

There are two key autumn periods:
— Period of autumn grass build-up
— Managing the final rotation

Generally, rotation length should be extended from 1 August. The focus of this period is to gradually build pre-grazing herbage mass, targeting maximum covers of 2,000 kg to 2,300 kg DM/ha in mid-September. Pre-grazing covers >2,500 kg DM/ ha are difficult to utilise and should be harvested as surplus (round bales). Surplus paddocks should be removed in August. Removing paddocks after the first week of September should be avoided if possible; a September harvest is too late as paddocks do not have enough time to re-grow to make any contribution in the last rotation. By achieving the right farm cover at the right time, decisions are easier to make. Many farmers fall into the trap of building cover too late and are pushed into harvesting excess grass in September.

Autumn nitrogen application

Grazing stocking rates vary considerably on farms and have a huge effect on feed demand. As the Nitrates Directive deadline date for nitrogen application is 15 September in Ireland, farmers must decide in late August/early September what level of nitrogen application they will apply to ensure sufficient grass growth for the final rotations. Farmers with a high grass demand in October/November who have their nitrogen applications updated by August should consider applying a blanket application by mid-September. The amount to be applied may vary and will depend on feed supply. Nitrogen should only be blanket spread if the farm is under target for grass. Spreading excess nitrogen in autumn is a waste of money as the soil is naturally releasing nitrogen.

Preparing for the final grazing rotation: August–December

The aim of this period is to maximise the amount of grass utilised from September to December while also finishing the grazing season with the desired farm grass cover. The farm grass cover or amount of grass grown over this period will depend on stocking rate, level of supplementation and autumn nitrogen application. The following guidelines should be followed:
— Rotation length should be increased from 25 to 30 days in mid- to late August to 35 to 40 days in late September.
— Last grazing rotation should be 30 to 40 days with first fields rested from 5 to 10 October.
— Closing should be a week to 10 days earlier on heavier type soils.
— Choose drier paddocks, paddocks close to the yard or sheltered paddocks to close first so that they will be the first ones grazed in spring.
— Close the wettest paddocks next in the rotation followed by the remaining paddocks.
— Aim to have 60% of the farm grazed by the first week of November with the remaining 40% grazed by late November/early December (these dates will change depending on location).

– Aim for an average farm cover of (1,000 kg DM/ha) by late September.

– Pre-grazing yields should not exceed 2,000 kg to 2,300 kg DM/ha. Very high pre-grazing yields will result in poor pasture quality and poor utilisation by the grazing animals.

– Avoid grazing very high pre-grazing yields in last rotation (>13 cm; >2,500 kg DM/ha). As well as the problems outlined above, high pre-grazing yields in the last rotation have been shown to have a very detrimental effect on subsequent winter/ spring grass growth.

– Pastures should be grazed well (3.5 cm) in the last rotation to encourage autumn/ winter tillering. Use younger or lighter animals or dry cows to achieve this residual in wetter conditions.

– The closing grass cover in early December should be, on average, 650 to 700 kg DM/ha.

Autumn 60:40 rotation planner

The autumn rotation planner is a tool to help extend the grazing season into late autumn and, if followed, will ensure that paddocks are set up correctly for grazing the following spring. The 60:40 plan is based on having proportions of the farm closed by certain dates. These dates will vary slightly across the country and depend on soil type and the amount of grass that is likely to grow over the winter. The 60:40 autumn rotation planner will not tell you if you are grazing paddocks that have too much grass or if you are not achieving desired post-grazing residuals. You will have to gauge that by walking through your paddocks or fields and assessing either visually or by measuring. The objectives of the autumn rotation planner are to:

– Keep grass in the diet of the animals for as long as possible.

– Set up paddocks for grazing the following spring.

The simple rules are:

– Dry farms: start closing between 5 and 10 October; 60% of the farm should be grazed by first week November; the remaining should be 40% grazed by 1 December.

– Heavy or slow grass growing farms: start closing on 1 October; 60% of the farm should be grazed by 20 October; the remaining should be 40% grazed by mid-November.

Figure 2.8 shows the difference between a dry farm (closing 10 October) and a heavy farm (closing 1 October). For a dry farm, 60% should be grazed within four weeks

and the remaining 40% in the next four weeks. On a wetter farm, this is adapted to 60% grazed in four weeks and the remaining 40% in three weeks. Over time, groups of animals can be housed, reducing the number of animals at grass.

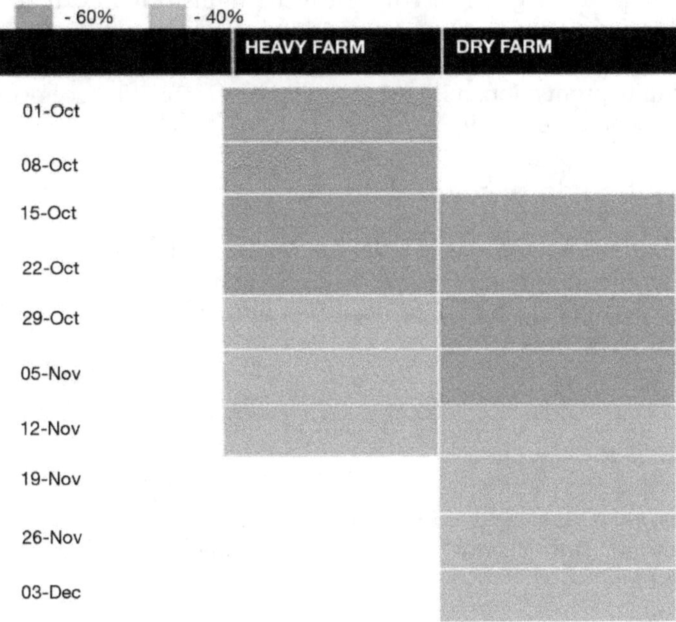

Figure 2.8. Example of an autumn rotation planner.

Managing a wet autumn

– Maintain a flexible attitude and do not allow poaching.
– Use the most sheltered and driest paddocks when grazing in wet weather.
– Strip grazing or block grazing can be used during wet weather to ensure minimal damage from poaching. Use one section per day to get the most from the grass.
– Where possible, use a back fence. This will help to protect re-growth and prevent soil damage.
– On/off grazing can be practised to reduce poaching damage and keep animals at grass for longer.
– Have multiple access points into a paddock so that grazing animals do not have to use the same entrance. If this is not possible, create a four- to five-foot grass roadway on a fence line to get animals to the back of the paddock.
– Strategically place water troughs in the paddock so that they will service several strips or blocks when a strip wire is being used.
– Graze paddocks with heavy covers from the back of the paddock on the sheltered side.
– Change the grazing break daily or every two days.
– Change animals at the same time; give them a routine.

Dairy

The main aim during this period is to maximise the amount of grazed grass in the dairy cow's diet. Supplementation may be necessary in certain situations, such as to:
– Reduce demand for grass so covers can be built to help extend the grazing season
– Maintain milk lactose levels
– Maintain milk production; milk prices will have a large influence over this
– Increase cow BCS before drying off

Supplementing dairy cows in autumn

If grazing full-time, 2 to 3 kg DM of a high energy low-protein concentrate is sufficient. If cows are indoors on grass silage (68 to 70 dry matter demand – DMD), 3 to 4 kg DM of an 18% crude protein concentrate will support production of 10 to 12 litres of milk (0.8 to1.1 kg milk solids).

Sheep

– Grass year starts in autumn
– Closing from late October
– Swards should be rested for 120 days over winter
– Match autumn closing dates to expected lambing dates to match spring grass supply to demand

▸▸ Farm infrastructure

F. Bogue, M. O'Donovan, L. Bastiaansen-Aantjes, A. van den Pol-van Dasselaar

The design and layout of grazing infrastructure is crucial to overall herd performance as it can allow more days at grass and hence greater profitability. It comprises three factors:
– Paddock system
– Road system
– Drinking water infrastructure

How to create an efficient paddock system

– Get a map of the farm with areas for each field/paddock.
– Decide on the number of paddocks required; this will depend on whether the paddock will be used for one, two, three or more grazings.
– Determine the most suitable road layout to service each paddock.
– Determine the most appropriate water trough position(s) in each paddock.
– Allow for multiple entrances into each paddock.
– Ideally keep paddocks square/rectangular; ideally, the depth: width ratio should be 2:1.

Key risks with respect to paddock layout

– Long narrow paddocks: too much walking over ground to graze the end of the paddocks can result in an excessive risk of poaching.
– Large paddocks: grass regrowth is grazed if over three to four grazings per paddock are needed. Using a strip wire to divide the paddock requires extra labour during the main grazing season and reduces milk solids.
– Small paddocks: insufficient grass for one grazing, extra water troughs are required.
– Farmers expanding their operation should use strip wire until they decide how many cows they will milk.

Alternatives to a paddock system

If there is no set paddock system, temporary wire can be used for all grazings. The advantage to doing this is that the grazing area can be adjusted throughout the year and that surplus grass/silage is more easily harvested.

Setting up the farm for grazing

– Get a farm map with exact areas of each paddock.
– Number every paddock.
– Create a good network of roadways.
– Roadways should follow land contours where there are extreme land features, and be wide with gentle sweeping bends (especially for larger herds).
– Locate roadways on the sunny, windy side of a ditch, hedge or tree line.
– Avoid putting races directly through springs or swampy ground.
– Plan underpasses carefully to allow for gentle slopes into and out of the underpass for drainage.
– Create multiple access points where possible to help with grazing during wet weather.
– Have several gateways between adjacent paddocks.
– It is a good idea to have easy access cut-off switches on electric fences.
– Have multiple water troughs or fittings where water troughs can be installed (this allows flexibility when putting up strip wires).
– Keep a record of dates when grazed, fertilised, topped and cut for round bale silage.
– Know the reseeding history and soil fertility of each paddock.
– Record varieties sown when reseeding.
– Test soil every two years.
– In drystock systems, assign specific paddocks to stock, i.e. cow paddocks, fattening stock paddocks, leader/follower paddocks.
– Maintaining a small number of grazing groups will allow the total number of paddocks required to be maintained at a manageable level. This can be done by grazing steers and bulls together and by mixed grazing of cattle and sheep and leader/follower systems

Paddock sizes

Proper subdivision of grazing land into paddocks is essential to be able to successfully manage pastures and achieve desirable rotation intervals. Paddocks must be connected with an efficient roadway system so that the herd can move easily. The ideal paddock system should include:
– About 20 to 23 full-sized paddocks and a few small paddocks near the parlour/sheds.
– The roadways/lanes from the parlour/farmyard to the paddocks should be wide smooth and as short a distance as is practical.
– The paddocks should be big enough so that there is sufficient pasture for the full herd for 24 to 36 hours when the pre-grazing cover does not exceed 1400 to 1600 kg DM/ha and on a 21-day grazing rotation.
– Paddocks should be rectangular to square in shape and wetter paddocks should have longest sides running adjacent to the races to avoid poaching in wet weather.
– Alter paddock shape to facilitate stock movement into and out of the paddock (i.e. stock move downhill to exit paddocks).
– Main paddock gateways should be angled to the roadway with at least two gateways for each paddock.
– The paddocks should be numbered with a tag on the gate and on a map of the farm.

Creating paddocks

1) Use the maps to consider several different ways of laying out the farm and consider the positives and negatives of each one.

2) Chose the option which corresponds to the most positives and the least negatives.

3) Mark the layout on the ground with marker pegs. Use different colours for roadway edges and paddock boundaries.

4) Re-consider the layout both from the practicality of construction and operation and from the perspective of the animal.
– Does it actually make sense?
– Are the paddock entrances on dry ground?
– Are the paddock entrances in the downhill corner of the paddock?
– Is the slope of the roadway less than 10%?
– Will the race disrupt the normal flow of water down a slope?

5) Re-align the markers on the ground to correct for the issues identified in no. 4.

6) Record the final layout on an accurate map and make several copies. Have a very large one made that is suitable to put on a wall.

7) In beef cattle systems, the ideal size for a 40-cow suckler herd is 2 ha/paddock.

Sheep

Using a paddock system/rotational grazing gives farmers more control, which leads to increased grass utilisation, growth and quality.

– Rotational grazing: minimum of 5 paddocks
– Provide high-quality leafy grass
– Average residency of 5 days per paddock

At a practical level:
– Average 100 ewe flock SR 10 ewes/ha (4 ewes/ac)
– 5 grazing divisions of 2 ha (5 ac) each
– Can be permanently fenced and each split

Or
– Fence set up to allow five to 10 temporary divisions as needed
– Allows flexibility and control

Why divide paddocks into 2 ha areas?
– Example: In week 9 of lactation (mid-May), the daily requirement for ewes is 3 kg DM/hd/day. Lambs require 0.7 kg DM/hd/day. $3 + 0.7 \times 1.6 = 4.12$ kg DM/ewe and lamb unit/day. $4.12 \times 100 = 412$ kg DM/ha/day = flock demand
– Pre-grazing yield = 1200 kg DM/ha $\times$ 2ha = 2400 kg DM available
– 2400 kg/412 kg = 5.8 days to graze paddock
– Paddock cover and flock demand will vary, which makes 2 ha per 100 ewes a good guideline.
– Ideally, this would be offered one hectare at a time with two to three days grazing to increase utilisation and improve regrowth potential.

Water

A water supply in each grazing division is necessary. Ideally, every paddock should have a permanent water supply. Placing troughs across fences reduces the number required. If using a temporary wire to strip or block graze, strategically place troughs in the field so that animals do not have to walk back over the grazed area for water. Alternatively have multiple fittings where water troughs can be installed (this allows flexibility when putting up strip wires). Water systems should:
– Deliver sufficient water to meet the stock needs during greatest demand. This means that the amount of storage and flow rate should be adequate, so that any trough is never less than two-thirds full.
– Use taps in easy to find, key locations to split the water system into sections so that it is quick and easy to shut off sections of the farm.
– Make it easy to identify, locate, isolate and repair leaks, using easily visible storage.
– Keeping water troughs in the centre of the paddock allows for livestock to be further split with temporary fencing.
– Alternatively, water troughs can be fitted with a long length of water piping and the water trough can be moved between grazing areas within a single paddock.
– Water supply/pressure will often dictate the size and type of the water trough used.

Farm roadways

Roadways are an obvious advantage as they allow easy access to paddocks and prevent soil damage. The aim is to have animals walking comfortably at 3 km/hr with their heads down so that they can see where they are placing their front feet (the back feet

will step into the same place). Actual animal walking speeds are determined by the walking surface, animal training and animal fitness. For roadway maintenance, keep the following in mind:
— It is essential that no water is allowed to pool on the race surface.
— Fill potholes as soon as they form.
— Remove any build-up of material at the sides of the race that will prevent water running off.
— Restrict the speed of tractors, quads and other farm vehicles on farm roadways.
— Keep vegetation trimmed well back to allow in light and wind onto farm roadways.

Temporary fencing

Temporary electric fencing should be used to divide larger fields to give the required paddock size, especially when grazing silage fields during the first rotation.

Chapter 3

Hay and silage making

Giovanni Peratoner, Alain Peeters and Riccardo Negrini

For Atlantic and Nordic Europeans regions, stocks are needed for winter seasons. For other regions, with drier water regimes and less predictable herbage growth, stocks are essential.

The scenarios related to climate change include the probability of increasing variation among years for temperature and rainfall regimes. It implies that the stocks will be essential for the resilience of the farming systems.

The stocks may be either hay or silage depending on the local weather conditions.

▸▸ Cutting management and haymaking

G. Peratoner and A. Peeters

Cutting as a form of grassland management

By cutting grassland, the whole aboveground biomass accumulated during each growth cycle is removed at once above a certain cutting height. Contrary to pastures, animal-mediated factors such as selectivity, trampling and deposition of excreta no longer affect the vegetation. Manure from the cattle can be spread evenly on the fields. Depending on the method used to mow the plants, a certain mechanical disturbance of the soil due to the machines and implements used is caused. Unless only small amounts of forage are harvested and immediately fed to the animals, the harvested herbage must be conserved. This means that it must be brought as soon as possible to a stable state, thus allowing preservation, to minimise losses in nutrients and quality during the conservation process.

Cutting height

A cutting height between 5 cm and 7 cm should be targeted. This cutting height allows the maintenance of critical amounts of plant tissues capable of photosynthesis for regrowth. At lower cutting heights, plants have to mobilise considerable amounts of nutrient stocks from their root systems. This leads to higher sensitivity to drought, delayed regrowth and lower yield at the following cut. Cutting too close to soil surface on an irregular soil surface may lead to scalped swards, which impacts vegetation, damages equipment and can lead to soil contamination of herbage. Moreover, for haymaking, if the drying process takes place on the field, excessively short stubble would fail to maintain distance between the overlying swath and the soil, causing poor air circulation below the swath and thus a delay in drying. A further increase in cutting height (i.e. to 10 cm), on the other hand, has been shown to not improve quality and decreases forage yield. Avoiding soil contamination during harvest is a pivotal issue, relevant to all forms of forage conservation. This is achieved by driving carefully, taking soil conditions under consideration (avoiding mechanical damage of the sward, which is difficult on steep slopes and under wet conditions) and setting the right cutting height (to be set on solid ground) in order to avoid soil scalping. The right working height (3 cm on solid ground) of all other implements to be used (conditioners, tedders, windrowers, loaders) must be accurately set as well.

Harvesting time

As the plant phenology stages progress, and especially with the start of the generative stage, there is an increase in forage yield and a concomitant decrease in forage quality. The latter is due to both a proportional increase of fibre-rich tissues (typically stems) and the lignification of other tissues due to senescence processes. While maximum yield is achieved at a later stage (after flowering), forage quality is low since it moves in the opposite direction to forage yield. The optimum harvesting time is therefore a compromise between forage yield and quality. Choosing the correct harvesting time is especially important at the first cut, which usually contributes most to the annual yield (Table 3.1). The duration of the growing season, the harvesting time at the first cut and the time needed for regrowth to accumulate enough herbage for the next cuts (usually five to eight weeks) determine the cutting frequency over the whole growing season. At cutting frequencies between two and four cuts per year, the contribution of the first cut varies between about 40% and 60% and decreases with increasing cut frequency.

Moreover, it is at the first cut that the proportion of reproductive stems is the highest, especially in systems with a limited number of annual cuts. This results both in a fast yield increase and a corresponding quality decrease (see example for mountain meadows in Figure 3.1). These changes in forage quality over time are less pronounced for regrowth (see example in Figure 3.2). Even with a clear choice of optimum cutting period, unfavourable weather conditions can cause considerable deviations of the harvesting time from the optimum. For this reason, conservation methods reducing the dependency on weather results such as silage and haylage can be alternatives to haymaking.

Table 3.1. Example of contributions of different cuts to the annual yield. Sources: www. gruenland-online.de for 5- and 6-cut meadows, and means of a 10-year series at two experimental mountain sites in South Tyrol for the other cut frequencies (Laimburg Research Centre, unpublished data).

Cutting frequency (cut/year)	Cut	Contribution to yearly yield (%)
2	1st	62
	2nd	38
3	1st	47
	2nd	28
	3rd	26
4	1st	39
	2nd	20
	3rd	21
	4th	21
5	1st	30
	2nd	20
	3rd	20
	4th	17
	5th	13
6	1st	30
	2nd	25
	3rd	15
	4th	15
	5th	10
	6th	5

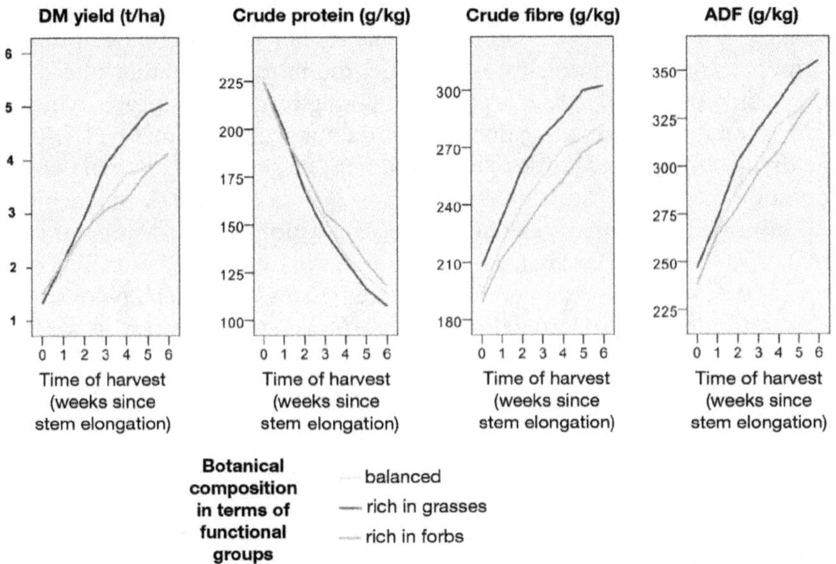

Figure 3.1. Change over time of forage yield and quality during the first growth cycle of mountain meadows. Means of 202 environments (site × year) in South Tyrol (north-eastern Italy) at altitudes between 660 and 1660 m asl (Laimburg Research Centre, unpublished data).

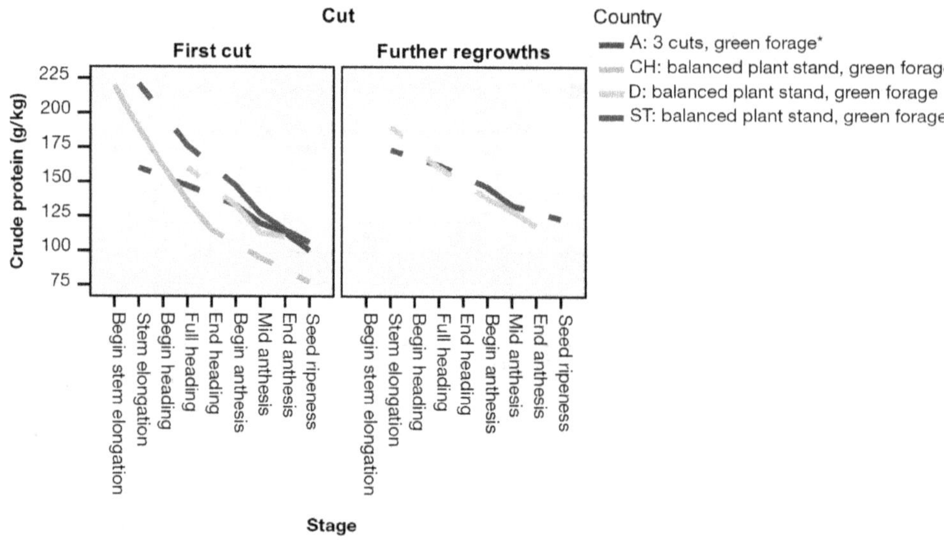

Figure 3.2. Changes in crude protein content depending on the average phenological stage of grasslands according to tabulated forage values of four Alpine countries (adapted from Resch *et al.*, 2006, Daccord *et al.*, 2007 and DLG, 1997). * Plant stage estimation according to crude fibre content.

Haymaking

Haymaking is a physical process based on drying herbage in order to produce hay. The aim is to reduce, as quickly as possible, the moisture content of the herbage from around 70% to 80% to about 13%, making it safe for storage, which means that organic matter degradation due to bacteria, fungi and enzymes no longer occurs. Rapid drying reduces losses due to respiration processes (oxidisation of plant sugars, degradation of protein into amino-acids) and decreases the risk of leaching losses due to wetting of the crop by rainfall. Plant respiration is almost stopped at a water content of 30% or less. Rain induces quality losses in four ways: 1) leaching of soluble carbohydrates, vitamins and minerals, 2) increased and prolonged plant respiration, 3) leaf shattering, and 4) microbial breakdown of plant tissue. Typical losses in several nutrients in rained-on hay can reach about 10%.

Drying takes place, at least partly, on the field by the action of sunlight and wind. Low relative air humidity, combined with high temperatures and moderate breezes are favourable conditions for drying. Water losses are initially rapid, especially from leaf tissues. With increasing dry matter contents, the remaining water is increasingly tightly bound to plant tissues (especially in stems). After leaf wilting, the stomata close and the water flow from stems to leaves is interrupted. Leafy species, and in general forbs and legumes, have on average higher water content than grasses and require longer drying times. Equipping mowers with conditioners, causing abrasion, bruising or crushing of plant tissues by a mechanical action immediately after mowing, speeds up wilting. Following the cut, a curing phase takes place in the field. The crop should be first tedded (once) and turned (twice) on the first day, then all further operations should be performed very carefully (low round per minute of rotating implements) to reduce crumbling losses, a risk to which the leaves are particularly exposed. Crumbling losses can play a pivotal role in determining the final forage quality, as leaves represent the most valuable plant parts. Crumbling losses increase with increasing dry matter content of the herbage being tedded, turned or swathed and with its leaf-to-stem ratio. Swathing is usually performed just before collecting the herbage to be transported to the barn, but it may also be advisable if there is a risk for the herbage of getting wet. This includes rainfall as well as overnight dew, which is likely to occur in autumn. If crumbling losses are not a pivotal issue and stable weather is foreseen, a less labour-intensive strategy can be used, such as tedding once on the first day (if there is no conditioner), then tedding or turning 48 hours after cutting and swathing after 72 hours before baling on the fourth day. When weather conditions are not favourable, it is preferable to conserve forage as haylage, rather than risking producing low quality hay.

Legumes and legume/grass mixtures must be wilted more carefully than pure grass hay. Leaf losses can indeed be much more important in alfalfa and red clover based swards compared to grasses. Hay must be tedded at lower speed and early in the morning when dew makes the forage leaves supple. Later in the day, forage leaves are too dry and breakable, which can induce important losses and thus nutritive quality reduction.

Some species are better adapted to haymaking than others. In grasses, cocksfoot, tall fescue or timothy can be dried more easily than ryegrasses. In legumes, lucerne can be wilted more easily than red clover.

Effect of topography

Depending on slope, different machines can be used for harvesting, curing and collection operations. Up to a slope of about 40% to 50%, harvest operations can be fully mechanised (i.e. for mowing, tractor-mounted disc mowers and specialised two-axle mowers can be used), but beyond such slopes, implements requiring more and more human labour are required (bar mower or even manual mowing using a scythe). This results in higher costs in both labour and machineries, almost doubling from flat on slightly sloped areas to very steep areas with slopes beyond 60% (Figure 3.3).

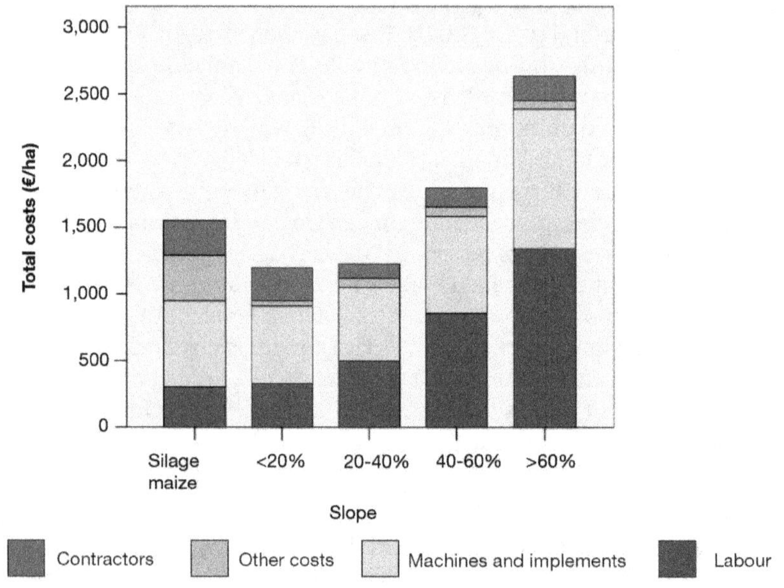

Figure 3.3. Production costs of forage depending on slope (Peratoner *et al.*, 2015, modified).

Methods of hay drying

Field drying

The whole drying process takes place in the field. The advantage of this method is the lower requirements in terms of equipment and thus in low capital expenditures. The main disadvantage is the long period of time with good weather required to achieve forage stability. Three to five days with favourable weather, depending on the harvested material, the specific weather conditions and the latitude in Europe are usually necessary to achieve a safe state. This results in a high risk during periods with unpredictable weather. When there is unusually frequent rainfall, the choice of a suitable harvest time may become very difficult and close attention to weather forecasts must be paid. In mountain regions, in which orographic rainfall cannot be accurately forecasted, the prediction of a four-day harvest window cannot be taken for granted. Moreover, getting the grass to a safe state requires a higher number of curing operations at increasing dry matter contents of the herbage, resulting in a high risk of crumbling losses.

Barn drying

Herbage is dried down to a moisture content of 25% to 40% (depending on the performance of the drying facility) and then further dried and brought to safe state in the barn. Several solutions are available to provide air movement and/or increase air temperature or even to decrease relative air humidity and thus to increase the drying efficiency (roof collectors, photovoltaic energy, dehumidifiers). The main advantage of this method resides in the shorter time (one to two days) needed for the field phase, and therefore in the lower risk due to weather unpredictability. Moreover, the reduced need for curing operations in the field of herbage with high dry matter content leads to a reduction of the risk of crumbling losses. The main disadvantage consists in the capital investments needed for the drying facility and the crane, which is needed to efficiently move the forage to be dried within the barn. Adequate capacity for average yield is needed.

▸▸ Silage

A. Peeters and R. Negrini

Silage developed considerably in Europe in the 1960s as an alternative to hay to better conserve high-quality forages, especially in rainy climates.

Silage can be made with younger forage than hay and thus with better quality forage. Grass is cut and wilted in the same way as for hay but for a shorter period of time. Consequently, the harvesting process is less dependent on weather conditions. Silage is usually prepared after a short wilting period of one or two days, bringing grass dry matter content from about 20% to ideally around 25% to 30%.

The optimum cutting stage for silage making is between the optimum grazing and haymaking stages. In spring, it corresponds to the stem elongation stage. Vegetative grass can be ensiled too. A compromise must be found between forage quality and costs per tonne of DM harvested. Early cuts are of high quality but cost more per tonne harvested than later cuts. Weather also remains a constraint. When the ideal cutting stage is approaching, the cut has to be taken when weather conditions are favourable.

Silage making is an anaerobic process. Forage is compressed to eliminate oxygen as much as possible. This can be done in different silage devices.

After the wilting process, grass can be collected by self-loading wagons and transported loose and stored on a concrete floor surrounded by concrete walls, with the entrance of this structure remaining open on one or two sides for loading grass. Tractors run over the green forage after each layer deposition to compact it. Finally, when the bunker is full, silage is covered by a plastic sheath to prevent oxygen coming back into the silage and to protect it from rainfall and sunlight.

The ensiling process (Figure 3.4) is then started. There are six different phases.

Phase 1

Aerobic microorganisms present on the grass leaf surface consume the oxygen of the forage mass to create the desirable anaerobic conditions. Unfortunately, aerobic respiration also consumes water-soluble carbohydrates needed by the beneficial lactic acid bacteria in a later step. All the oxygen is eventually consumed. This process generates a temperature increase, carbon dioxide emission and thus nutrient losses, and water evaporation. Under good conditions, this phase lasts only a few hours.

Phase 2

In the beginning of this phase, oxygen is depleted. Anaerobic bacteria replace aerobic bacteria. The primary bacteria are *Enterobacteria*. They can tolerate heat and can thrive in a pH ranging from 7 to 5. These hetero-fermenters produce both acetic (a weak acid) and lactic (a strong acid) acids. The final proportions of these acids depend on grass maturity, moisture, and the composition of spontaneous bacteria communities. This phase usually lasts 24 to 72 hours.

Phase 3

When the pH is lower than 5, homo-fermenter bacteria take over. These bacteria are more efficient than hetero-fermenters. They produce lactic acid. Lactic acid is the most desirable of the anaerobic fermentation acids. Silage pH quickly decreases. As the temperature of the silage mass decreases and the pH continues to drop, bacteria are inhibited. Phase 3 is a short and transitional phase. It usually lasts only 24 hours.

Phase 4

Homo-fermentative bacteria continue converting water-soluble carbohydrates to lactic acid. The temperature stabilises. In well-preserved silage, lactic acid should represent more than 60% of the total silage organic acids. This lactic acid phase is the longest of the ensiling process. It continues until the pH is so low that all bacteria are killed. Fermentation stops and forage can be then conserved for a long period. The final pH depends on the type of forage and the dry matter content of the ensiled forage, and can range from 4 to 4.5.

These four first phases can be completed within 10 days to three weeks from harvest. It is thus recommended to wait at least three weeks before feeding freshly prepared silage.

Phase 5

The fermentation process has stopped and other changes occur. Starch becomes more quickly degraded in the rumen with longer storage times. Other changes may also occur in the digestibility of the neutral detergent fibre (NDF).

Phase 6

When feed out begins, oxygen is introduced into the silo. This can lead to substantial dry matter losses. Proper management of the silage face can minimise losses.

The grass silage process described above is similar to what happens in the preparation of sauerkraut (fermented cabbage) or pickles in brine.

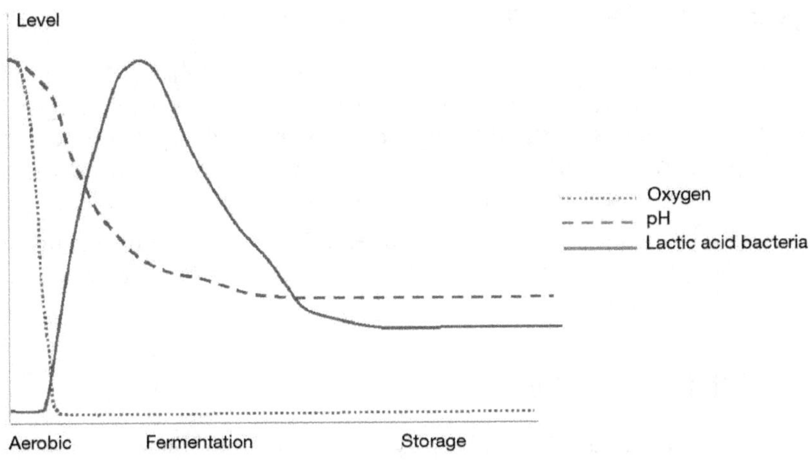

Figure 3.4. Oxygen, pH and lactic acid bacteria population evolution over time in a favourable ensiling process (after Dunière *et al.*, 2013).

There are also other ways to conserve silage in addition to a bunker silo. The fermentation process is the same.

Self-loading wagons can deliver harvested forage at the base of a silage tower. A blower then brings wilted forage through a pipe at the top of the tower and injects it inside. Compaction occurs naturally by the effect of forage weight. Forage material at the bottom of the tower should be ensiled at a high DM (35% to 40%) to prevent effluent release. The upper surface of the silo is not compacted, it is exposed to air. Consequently, aerobic fermentation can occur on the first meter, leading to forage losses.

A simple silo can be prepared on the ground, in a field. It is compressed by tractors and covered by a plastic tarp. This type of silage heap is cheap, but it can be contaminated by earth and the compaction work is dangerous because the tractor can turn over while running on the sloping side.

Wilted forage can be harvested and compressed by round or square balers. Bales are then packed in plastic wraps. This system is more expensive than the first two, but it offers flexibility in the silage making process. Small plots or plot sections can more easily be harvested because this system does not require to open a large silage bunker which could lead to silage degradation. Silage bales are also easy to distribute to livestock in barns or outdoors. Round bales can, for instance, easily roll on the ground which facilitates forage distribution. Bales should be checked regularly because crows and rodents can create holes in the plastic through which air can penetrate and cause forage quality degradation. Punctures should be quickly fixed. Bales should be stored on a compacted or stony floor to prevent rodents making tunnels underneath. Hedge bases are particularly unsuitable locations for storage.

Forage species effect

Legumes such as alfalfa and red clover have lower water-soluble carbohydrate contents and a higher buffering capacity than grasses, especially perennial ryegrass. The final pH of legume silos is thus slightly higher than grasses. Nevertheless, it is entirely possible to make good legume or grass/legume mixture silage when proper practices are adopted. A key issue is to get sufficiently high dry matter content.

Sainfoin, a forage legume species adapted to temperate and Mediterranean climates and to neutral to alkaline soils, has a high water-soluble carbohydrate content and could be suitable for silage.

Undesirable fermentation

If forage is contaminated by soil, dry matter content is too low and pH is not low enough to stabilise silage, then other anaerobic bacteria (*Clostridia*) cause butyric fermentation (Figure 3.5). They transform lactic acid into acetic and butyric acids, and protein into ammonia. This reduces silage intake by ruminants.

If prevention measures for harvesting clean forage and reaching an optimum DM and water-soluble carbohydrate content are not sufficient, additives can be used. These additives are added in the silo. They can be formic or sulphuric acids, bacterial inoculants or soluble sugars such as molasses.

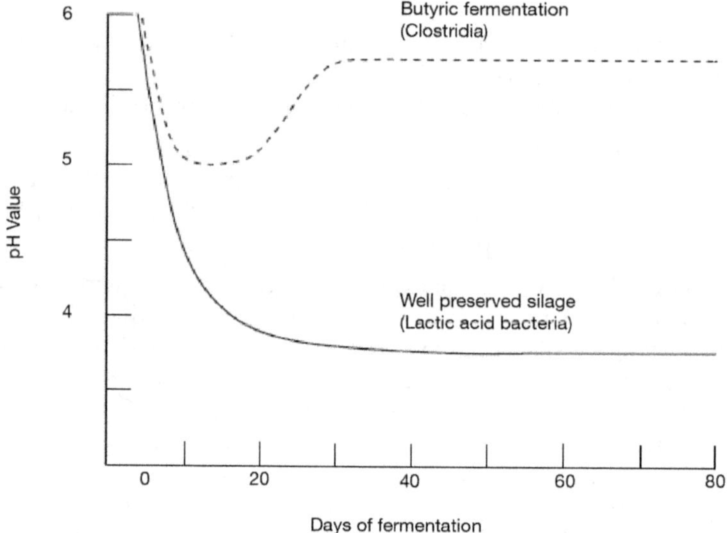

Figure 3.5. Changes in pH during silage preservation for butyric and lactic acid production. (Knabe *et al.*, 1986 in DairyNZ).

Loss prevention measures

There are a number of measures that can be taken to limit losses during harvesting and fermentation:
– Grass should ideally be cut after one or two sunny days to increase water-soluble carbohydrate (sugar) content. It can then be cut in the morning of the next sunny day. If weather conditions are less favourable in the preceding days, it should be cut in the early afternoon of a sunny day.
– Wilting grass should be done as quickly as possible to limit sugar losses. If possible, it should be done within 24 hours.
– In a bunker silo, silage should be well compacted with a heavy vehicle. For baled silage, a high-density baler should be used.
– The stack should be completely covered with a heavy, airtight cover. This cover should be first washed if not clean enough.
– A covered stack should not be reopened to add more forage later on.

Measures can be taken to limit losses during feeding out. When silage is exposed to air again, aerobic microorganisms can start using oxygen to digest silage nutrients. This causes a temperature increase and nutrient degradation. Losses can be limited by doing the following:
– Exposing silage to air should be limited as much as possible. This could be achieved by removing a section at least 20 cm deep each day on the front-facing side.
– Feeding out should thus be done every day, especially in summer.
– The stack face should be kept open on warm days to avoid heat under the cover.
– Cut silage off the face, rather than pulling it off. This keeps a smooth surface at the stack face, which reduces air penetration into the stack.

Silage quality can be checked by traditional nutrient value analysis and by specific fermentation quality analysis. This latter analysis includes:
– Ammonia nitrogen (NH_3-N) in per cent of crude protein (CP): proportion of N broken down during ensilage.
– Total fermentation acids (TFA): total amount of acid produced during fermentation. It includes lactic, butyric and acetic acids and possibly also propionic acid and ethanol.
– Volatile fatty acids (VFA): VFA is high when fermentation is poor.
– Lactic acid: typical lactic acid contents range from 60 to 150 g/kg, and higher values are better. A proportion of lactic acid in the total fermentation acids (TFA) > 70% is ideal.
– Acetic acid: high levels can restrict intake.
– Butyric acid: indicates poorly fermented silage.
– Residual sugar (RS): a valuable source of energy for rumen microorganisms.
– Ethanol: should be very low, because it is associated with the growth of undesirable yeasts.

Good quality silages are low in butyric and acetic acids, low in ammonia-N, and high in lactic acid and sugar.

⏩ Techniques adapted to legume and legume mixtures

A. Peeters and R. Negrini

Most scientific and technical publications on grassland forage conservation deal with pure grass mixtures, and especially pure perennial ryegrass. Publications on legumes are rarer. Those on mixtures are almost inexistent. However, farmers use mainly mixtures.

Mixtures harvested for forage conservation can include grass, legume and other dicotyledon species. Wilting dicotyledons in general requires techniques that are different than those used for grasses. This section focuses on legumes.

Seed mixture and sowing

Lucerne is often sown in mixture with grass. Cocksfoot is its most frequent companion grass. Lucerne/cocksfoot mixture is best sown from mid-August to mid-September. It will then produce a normal yield the following year. If sown in the spring, it is more likely to be invaded by weeds and the yield will be quite lower. Two cuts can be expected instead of four in a normal year.

Lucerne/grass mixtures have several benefits compared to pure lucerne. They cover the soil faster after sowing and can better control weeds. They stabilise yield repartition over time. For instance, in cold periods when lucerne growth is reduced, grass dominates the mixture and ensures a better yield. Mixtures are also easier to wilt and to conserve as hay, haylage or silage.

Cutting time

Typically, in Atlantic climates in north-western Europe, lucerne and lucerne mixtures can be harvested from mid-May onwards. They produce high DM yields without any nitrogen fertilisation. Average annual yield often reaches 15 t DM/ha in farm conditions. An example of the proportion of annual yield per cut is shown in Table 3.2.

Table 3.2. The proportion of annual yield per cut in a lucerne plot harvested four times a year.

Cutting number	Cutting time	Proportion of total yield (%)
1st	Late May	35
2nd	Early July	35
3rd	Mid-August	20
4th	Late October/early November	10

Source: Genever and McConnell, 2014.

Other slightly different timings for the four cuts are frequent: mid-May, mid- to end of June, end of July, end of September.

Table 3.3 presents average yield distribution collected in a farm network in the central-eastern part of France. These regions are drier compared to the regions in

Table 3.2. Average yields only reach 9 t DM/ha. In continental climates, yields of about 12 to 13 t DM/ha seem to be more typical.

Table 3.3. Yield distribution of lucerne (t DM/ha) in the three first cuts in central-eastern France (Drôme, Isère, Rhône and Loire).

	1st cut	2nd cut	3rd cut
Average 1985-1987 (t DM/ha)	4.0	2.6	1.6
% yield of each cut in total production	49	32	19

Average yield of the fourth cut: 1 t DM/ha. Results of a network of 43 farms (190 plots - 139 ha) (1984-1987) (Mauries, 1988).

Lucerne is cut at pre-bud stage because it is the best compromise between yield and quality, but it must be harvested at least once a year at full bloom to increase the lifespan of the crop (Figure 3.6). At pre-bud stage, flower buds can be felt under the fingers by grasping the end of a stem. On 20 stems picked at random, four should have flower buds. Figure 3.7 shows that root nutrient storage is only reconstituted at flowering stage.

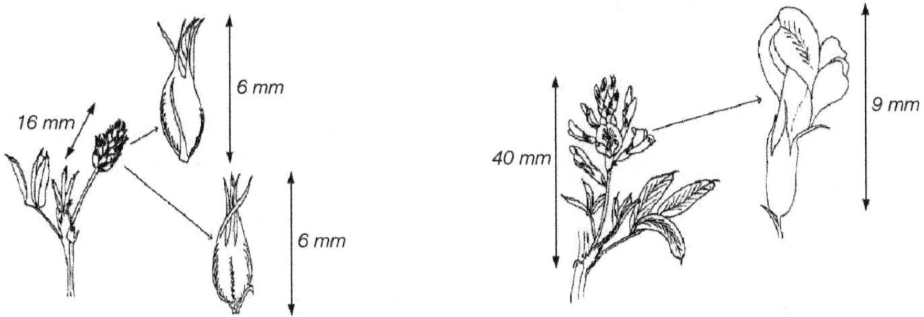

Figure 3.6. Bud (left) and full bloom (right) stages of lucerne.

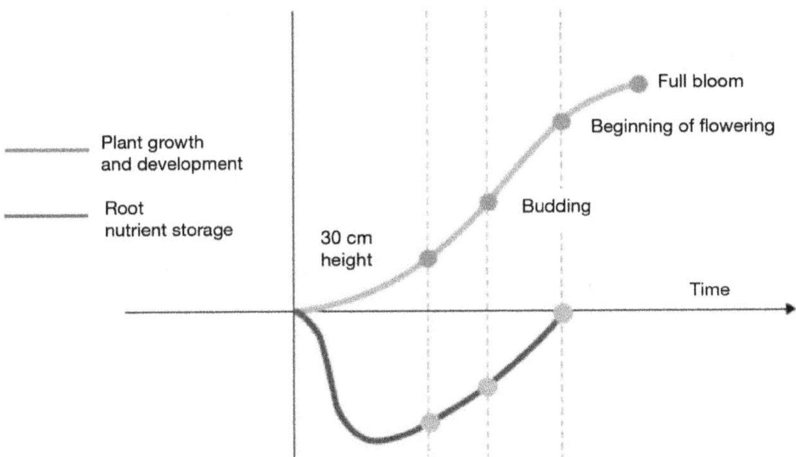

Figure 3.7. Evolution of physiological stages and root nutrient storage during an uninterrupted growth period of lucerne (Demarly in Delisle, 2010).

The last growth can be grazed to prolong the grazing period. This should be done by strip grazing with one-day occupation. Grazing is more interesting than harvesting the small available production at that time of the year.

Mowing and wilting

The cutting height is very important for persistence and total annual yield. Cutting too short reduces yield and persistency because plants have to use too many nutrient resources from their roots and crown to rebuild their leaf area after a cut.

The height of new stems is a good indicator to decide the best cutting time. New shoots should be visible but short enough to avoid cutting them with the mower (Figure 3.8).

Lucerne is almost always cut too short in practice. The minimum cutting height is 7 cm to avoid damaging the crown of the plant. This cutting height has two additional advantages. Cut forage is deposited on these relatively long stems which create a space between the ground and forage where air can circulate, which helps wilting the green mass. This cutting height reduces also earth contamination risk.

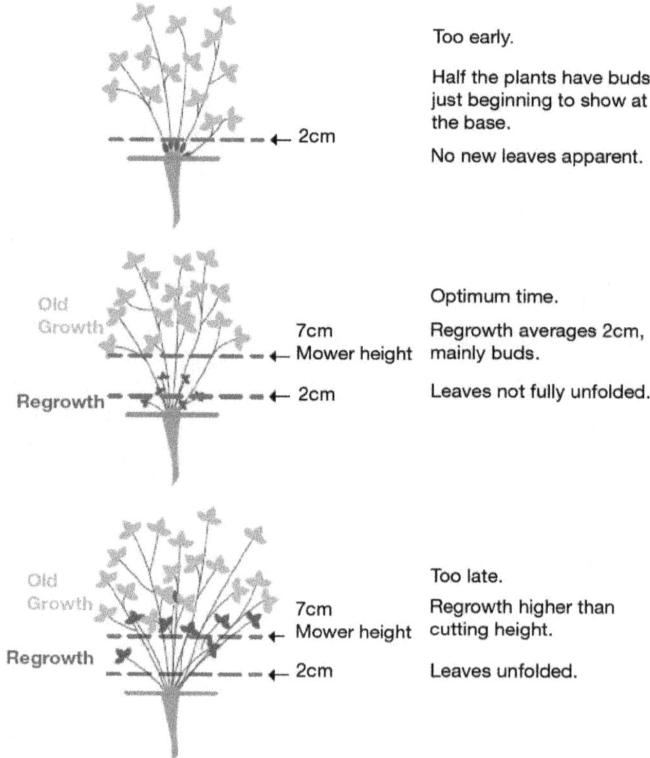

Figure 3.8. Determination of optimum cutting time of lucerne (Genever and McConnell, 2014).

Lucerne leaves can contain up to 70% of the protein and 90% of the minerals and vitamins of the aerial parts of the plant. Lucerne and other legumes such as red clover are difficult to wilt because the leaves – the most nutritious part of the plant –can easily detach from the stem and fall onto the soil if wilting is too aggressive. As wilting progresses this risk increases (Figure 3.9), and it is very high at the last passage of the mower for haymaking. The mower should run slowly, and mowing and raking the crop should be done in the morning dew just after sunrise. At that time, leaves are suppler. Tedders are responsible for most losses, while windrowers cause only about 3% to 5% of total losses.

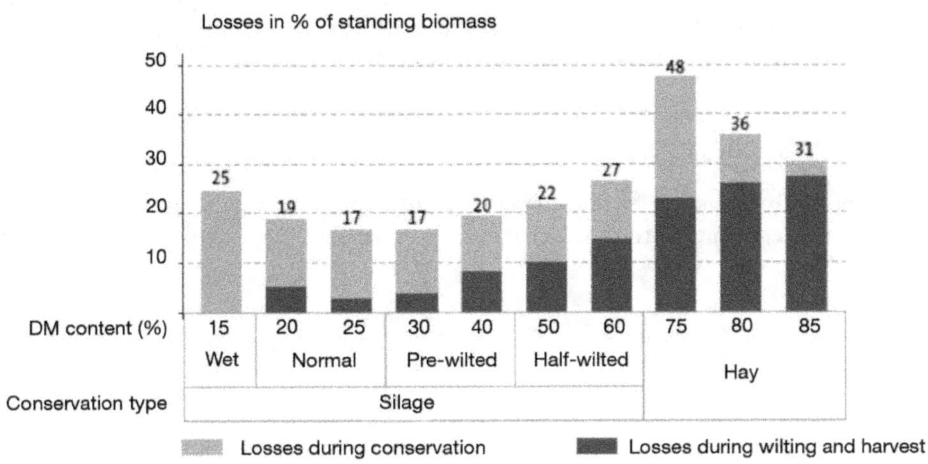

Figure 3.9. Losses at harvest and during conservation according to intensity of wilting and conservation type. Credit: Arvalis – Institut du Végétal in Delisle, 2010.

Lucerne/grass mixtures are easier to wilt because lucerne leaves and stems are diluted in a mass and protected by a weft of more flexible grass leaves. They are also easier to ensile due to the higher sugar content of grasses.

Roller-type mower conditioners can speed up the rate of moisture loss from the stem, but they can also increase leaf shatter. If adopted, they should be used with great care. Their cost is also a limiting factor for their use. The lucerne area should be big enough to recover the expense of purchasing this equipment.

Ensiling process

Lucerne silage can be preserved either in a bunker silo or in big bales. Target DM content should be 30% to 40% for bunker silage and 50% for big bale silage.

Lucerne stubble can pierce silage bale wraps. This should be prevented by using at least four layers of plastic. After moving bales from the field to the storage area, they should be checked for damage and, if needed, repaired.

Legume species

All the information described for lucerne is also valid for red clover, although red clover is more often mixed with ryegrasses, meadow fescue and timothy. Red clover/cocksfoot is also a very good mixture.

Red clover stems dry out more slowly than lucerne stems. This is not a problem for silage making, but it is for haymaking. Red clover is thus better adapted to silage. The high protein content (18–25%) of legume silages make them a good alternative to soybean meal in diets. This contributes to reduce the devastating effects of soy cropping in South America on species-rich habitats. In complement to maize silage, they bring protein and minerals that are lacking in this C4 grass. The lower readily available starch content and higher buffering capacity of lucerne compared to maize silage also has a beneficial effect on rumen pH. This may prevent rumen acidosis.

Nutritional characteristics

Lucerne and red clover have outstanding nutritional characteristics. The protein and calcium contents of red clover and lucerne are high. Forages of both species are highly digestible.

Chapter 4

Soil and nutrient management

Julien Fradin, Benoît Delaite, Xavier Delmon,
Martin Komainda, Giovanni Peratoner,
Nora Schiebenhöfer and Johannes Isselstein

The potential productivity of grasslands, like that of crops, depends primarily on climatic and soil factors. Soil and nutrient management can then increase (or reduce) this potential.

▸▸ Soil characteristics related to grasslands

J. Fradin, B. Delaite and X. Delmon

Soil is an essential resource (FAO, 2015) that is important to understand and use according to its specific characteristics. This is essential when looking at grassland production and ecosystem services provided by grasslands. The complexity of the soil lies in the fact that all its compartments are at first hidden.

What is soil?

Soil is an environment that is constantly evolving as a result of processes that alter the parent rock and transform organic matter. The key interest is in the part of the soil that may potentially be colonised by plant roots. It is generally accepted that it takes a century to form one centimetre of soil.

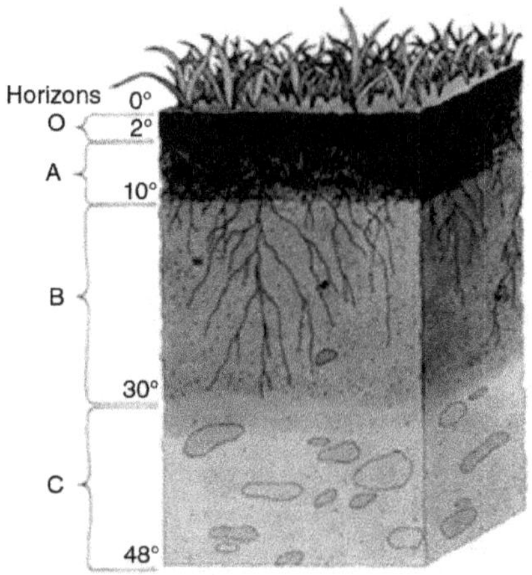

Figure 4.1. Soil composition. Source: NRCS – USDA.

O: Organic horizon where decaying matter accumulâtes; A: Topsoil or surface horizon, mixture of organic matter and minerals, highly fertile where most soil life occurs; B: Subsoil, little organic matter and accumulation of weathered materials (clay and oxides minérales); C: Substratum or parent material, layer that is not greatly affected by the weathering process.

Soil composition

Depending on its degree of compaction, soil porosity is around 40% to 60%, which refers to the amount of voids. This defines the volume of soil likely to be occupied by living organisms, roots, air or water. The solid part of soil is mainly made up of mineral matter, bringing total soil mass to about 95% (with the exception of peat soil). The remainder is organic matter, both living and dead. Although organic matter accounts for a very small amount of mass and is only concentrated in the first 20 centimetres of grassland soil, it plays a key role in its functioning. Soil contains the largest stock of organic carbon in the continental biosphere. The properties of organic matter have a strong impact on the environmental and agronomic problems affecting soil.

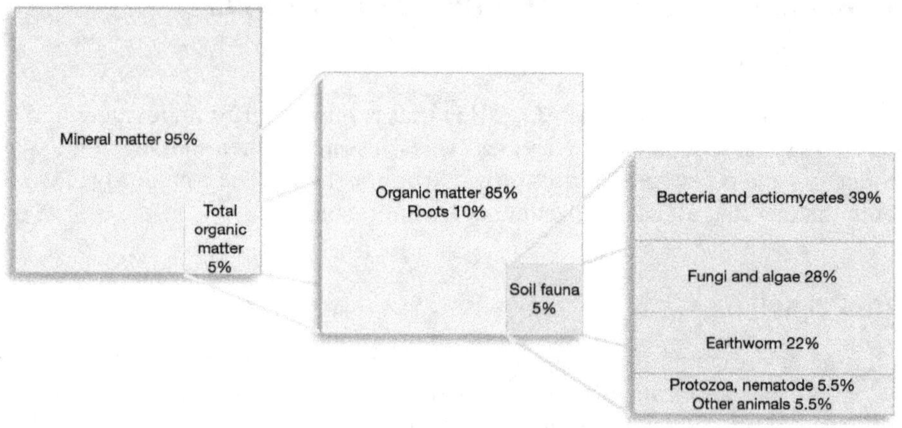

Figure 4.2. Topsoil layer components within temperate grassland (Bachelier, 1978).

Type and size of soil components

The parent rock determines some of the properties of soil, including its physico-chemical components. Soil particles are classified by size: from the biggest to the smallest, these are sand, silt and finally clay. The relative proportion of these types of materials determines soil texture, which provides a good idea of the general behaviour of the soil. Sandy soils offer excellent drainage and a good bearing capacity and enable grasslands to be farmed longer in the season; however, they also dry much faster than other types of soil.

The capacity of soil to store water is directly related to its texture, which reflects characteristics derived from geological events on which farmers have no influence. The rougher the particles, the more drainage capacity and the less water content. With the increase in soil organic matter content, which is generally high under grasslands, the amount of water stored increases significantly. The organic material acts as a sponge and can store up to 20 times its weight in water. But water stored in soil does not exactly match the water available to plants because it is too strongly bound to particles. Clay soils retain water better than any other type of soil, and partly prevent its use by plants. Loamy soils have the best capacity to store and supply water for crops.

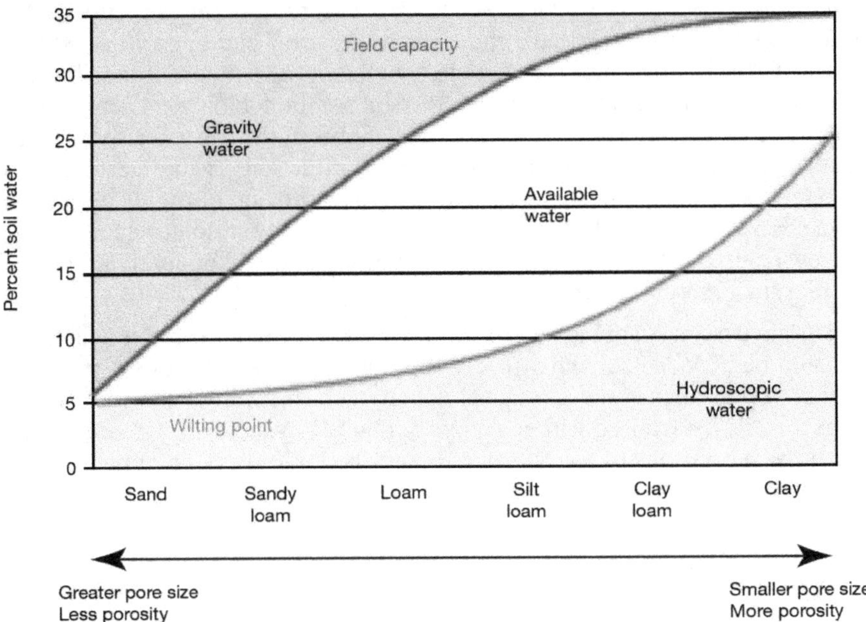

Figure 4.3. Soil moisture conditions for various soil textures.

Soil structure and physical fertility

The arrangement of soil particles, minerals and organic matter forms 'aggregates' of various sizes. Their juxtaposition confers the structure upon the soil. A good structure allows the penetration of roots and water while maintaining good soil cohesion. Soil tillage adds air to the soil, which makes rooting easier but reduces soil structure stability. Aggregates prevent water and wind erosion as well as reduce soil compaction. Strong biological activity allows the formation of a crumbly structure favourable to the flow of air and water. Earthworms are the organisms most involved in the formation of aggregates because they blend high volumes of soil in the case of grasslands. Soil structure stability increases with organic carbon content.

Soil fertility

Soil fertility is a concept that evolves over time and many definitions coexist. According to Abbott and Murphy (2003), soil fertility refers to soil's ability to provide the physical, chemical and biological conditions necessary for the proper development of cultivated plants. The concept of soil fertility can vary depending on the crop. Today, organic matter is considered as pillar of soil fertility.

Soil chemical fertility

The mineralogical clays have the ability to exchange cations (Ca^{2+}, K^+, Mg^{2+}, Na^+) thanks to their structure in the form of superimposed layers which are negatively charged. There are different types of clays that store mineral elements in varying degrees; the more the clays have the capacity to swell, the better their storing of minerals. Likewise, the finer the particles, the higher the theoretical soil fertility is because the exchange surfaces between the soil solution, biological activity and nutritive minerals are increased. The mineral elements of the soil can also be adsorbed by metal oxides (iron, aluminium) and organic matter. Soil organic matter is also the component with the greatest capacity to store and provide nutrients because of its sponge-shaped structure. In the case of a soil with a low cation exchange capacity, i.e., a low reserve of mineral elements, it is advisable to add mineral inputs in small quantities on a more frequent basis.

Minerals need a correct pH to be assimilated by plants. For good soil functioning, soil pH should be between 5.5 and 7 (6.2 to 6.7 is ideal). The H^+ protons are naturally released by plant roots and biological activity, which acidifies the soil. Biological activity in soils is favoured with a slightly acidic pH around 6.5. Organic matter in soil can fix the protons to a certain level, and the addition of carbonate via liming supplements this action.

Biological fertility

Soil is a biological ecosystem to the extent that it allows the development of living organisms. Grassland soils are home to the largest communities of macroscopic fauna; up to six or seven tonnes of organisms live in grassland soils.

These living organisms and decaying organic matter are a reservoir of nutrients. Biological activity is largely responsible for the mineralisation of organic matter. The size of the organic matter particles determines the type of fauna capable of degrading them with the enzymes they produce. The diversity of soil and microorganisms is therefore essential for the proper transformation of plant, animal and microbial material. The lower the C/N ratio of organic matter, the faster their mineralisation. Because of their constitution, plant residues, and especially roots, have the longest life in soils. The C/N ratio of organic matter in a soil analysis should be less than 10 to 12; any more than this and the organic matter will not be degraded due to a lack of available nitrogen.

There are two main types of organic matter: labile organic matter and stable organic matter. Labile organic matter is less than 10 years old and consists of large decaying residuals of 50 μm to 2 mm. It is the main source of energy for soil microorganisms. Through its development, soil fauna also provide the nutrients needed by plants. The more labile organic matter, the more biological activity and the more constituents are available for plant growth. Stable organic matter, also called humus, is on average 50 years old. It is not accessible by microorganisms because it is protected by physical or chemical barriers. When very fine and fixed to clays, it ensures the physical stability of the soil.

Grasslands are among the richest environments in organic matter. Given the prominence of organic matter for soil quality, it is essential to support them by avoiding ploughing of meadows too regularly and by having an integrated management approach to irrigation and fertilisation.

▸▸ Main macro-nutrients: N P K

J. Fradin, B. Delaite and X. Delmon

As previously discussed, the availability of mineral elements within the soil will impact grassland productivity. While the limiting factor is usually a macro-nutrient, most commonly nitrogen, soil and pH analysis should be used to determine the limiting factor.

Integrated fertilisation

Integrated fertilisation relies on three combined approaches:
— Assessing the elements exported and those available in the soil to calculate the necessary inputs.
— Calculating the elements available in the soil related to organic matter mineralisation and microbial fixation, particularly through legumes.
— Striking a balance between marginal productivity gains and marginal fertilisation costs; it is not economically viable to try to produce the maximum.

Nitrogen fertilisation

Nitrogen is the key element to monitor and the inputs are controlled because of risks of leaching. European legislation has established spreading standards depending on the form of nitrogen applied (mineral or organic), seasons, climate and slope.

> **A reminder on grassland N regulation**
>
> A maximum of 350 kg of total nitrogen/ha/year, of which a maximum of 170 kg of nitrogen of organic origin (including restitution by dung). This general regulation has been amended in several countries or regions (increasing to a maximum of 250 kg/ha in Ireland or parts of the Netherlands, or 230 kg/ha in Belgium, Austria, Denmark, Germany and parts of the Netherlands etc.). Special situations (watercourse, relief, soil type) may require lower levels.

Nitrogen exports depend on the type of grassland use (grazing, haymaking, silage making). There is a strong correlation between the soil organic matter content and the amount of nitrogen released by its mineralisation. This correlation is influenced by climate and soil. Beyond a certain threshold, there is no longer any correlation between the content and the supply of nitrogen. Manure patches in pasture meadows provide nitrogen as well. In Belgium, for example, it is considered that 100 LSU per ha per day return the equivalent of 9 kg N per ha.

Fertilisation will add nutrients to the soil. Macro- (and micro-) nutrients must first be returned by manure (taking into account regulations). Mineral fertilisers are only used as potential supplements. To calculate the inputs of farm manure, the following must be considered:
– The contents of elements in manure or slurry.
– The actual fertiliser value of the element, i.e. the quantity that will be released during the mineralisation throughout the year. This is expressed by a coefficient of equivalence with respect to mineral fertilisers.

For these calculations, different countries have tables that express the content of elements in different farm fertilisers, as well as their equivalence coefficient compared to mineral fertilisers. Equivalence coefficients take into account that organic molecules do not release fertilisers as quickly as synthetic molecules. On the other hand, organic fertilisers contribute much more to the soil organic matter content.

Ten rules for the proper application of farm manure (Agraost, Belgium)

– Know the fertiliser value of farm manures.
– Homogenise the product (mixing or dilution of manure, composting of manure).
– Ensure the quality of the product distribution behind the spreader: check the spreaders and slurry drums as well as the spread quantities. They should not exceed 15 cubic metres of slurry or 30 to 40 tonnes of cattle manure per pass. Thoroughly crumble the manure to limit the risk of smearing the harvested forage and the development of butyric bacteria in silage, as well as the appearance of voids in the grass, which will facilitate weed growth.
– Spread during favourable climate conditions: rainy or cloudy weather, with little wind and low temperatures.
– Work on soil bearing and short grass.
– Respect the needs of meadows.

– Spread during periods of optimum recovery, in line with good farming practices (nitrogen management plan) and limiting environmental risks. In general, the best period is from February to April depending on the region.
– In pasture meadows, avoid soiling the grass. Avoid the use of fresh manure, which leads to a reduction in palatability and therefore to poor use of the grass with the appearance of refusal areas. In addition, manure can promote the dispersal of certain pathogenic germs (Salmonella, Botula). These problems can be avoided by using compost, digestate or manure, making sure to work with ground injection systems (drip hose spreading boom).
– Limit volatilisation losses when spreading slurry by working as close to soil as possible or by injecting into soil.
– Respect the neighbourhood.

Nitrogen efficiency coefficient for supplement calculations

To finalise the calculation of supplemental synthetic fertiliser inputs, the following are deducted from the exported quantity*: soil contributions*, restitutions in pasture (dung), leguminous plant contributions and farm fertiliser (manure) contributions.

The quantities marked with an * are reduced by a coefficient of efficiency which takes into account that some of the nitrogen is not directly used for the production of fodder and goes into the soil, water, air or non-exported organic materials. This factor is usually between 0.7 (France) and 0.8 (Belgium).

Supply of phosphorus and potassium

The same reasoning for nitrogen applies to phosphorus and potassium. The ideal situation is to start from a soil analysis. The needs will be higher as grassland use intensifies.

In general, fertiliser inputs from farm manure (slurry, manure, compost manure) are sufficient and should not be offset by mineral fertilisers. Quantities can be estimated based on national farm fertiliser composition tables, taking into account (as with nitrogen) an equivalence coefficient that includes the mineral fraction not used by plants.

▶▶ Biological nitrogen fixation

J. Fradin, B. Delaite and X. Delmon

Among macro-nutrients, nitrogen is special because it can be provided by legumes that fix nitrogen from the air through symbiotic bacteria that live in nodules attached to their roots. This nitrogen is then available for both the legume and nearby plants.

In Belgium, comparative trials over five years have shown that, on average, a perennial ryegrass grown with red clover has a higher energy production and total nitrogen content than perennial ryegrass alone, even when fertilised with 400 units of nitrogen (Deprez *et al.*, 2005).

In addition to nitrogen fertiliser reduction, the grass-legume combination offers many other benefits (Fourrages Mieux, Belgium):
− Maintaining a dense and closed grass cover
− More stable production, especially in summer
− Better staggering of production periods
− Better palatability
− Better quality of fodder
− A good balance between mineral elements
− Better soil use through different root systems
− Forage quality that is maintained over time thanks to legumes

The main disadvantages of mixes are:
− More difficult control of harmful weeds
− Difficult management of the proportion of legumes in the mix
− Risk of greater losses during haymaking (loss of legume leaves)

Nitrogen intake according to the proportion of legumes

The proportion of legumes and grassland productivity must be taken into account to adapt nitrogen input (Source: GREN Bretagne, 2017). Nitrogen intake should be capped at the amount actually used by the plants over a growing season. This depends on the duration of growth was well as the type of soil, and therefore on productivity. In France, a maximum of 50 kg of effective nitrogen per hectare is generally taken into account.

Depending on the share of legumes in the mix (given as a percentage, estimated in the spring) and the grassland productivity (in t DM/ha), the nitrogen supply varies from 0% to 100%.
− Below 10% (visually, grasses largely dominate), the contribution of legumes is negligible.
− Between 10% and 30% of cover (visually, grasses ares dominant, but legumes are apparent), legumes will provide an average of 40% to 95% (white clover) and 30% to 75% (other legumes) of the grassland's nitrogen needs. This contribution steadily increases according to the grassland productivity (from 5 to 12 t DM/ha). See Table 4.1 below as an example for Brittany.
− Over 30% of legumes (visually, they can be seen almost everywhere), the nitrogen supplied covers all of the grassland nitrogen needs.

Table 4.1. Share of legumes contribution toward total nitrogen intake in mixed-grassland.

Grassland production (t DM /ha)	Share of nitrogen contribution for 10% to 30% of legume coverage
5	40%* – 30%
6	50%* – 40%
7	55%* – 45%
8	65%* – 50%
9	70%* – 55%
10	80%* – 60%
11	87%* – 67%
12	95%* – 75%

Source: GREN Bretagne, 2017. * The percentages followed by an asterisk correspond to white clover (*Trifolium repens*), the lower percentages correspond to other grassland legumes.

Additional support

A more detailed analysis is recommended to calculate the amount of nitrogen input needed. This will depend on the balance between exported nitrogen (which is based on productivity and crop maturity stage) and the nitrogen supplied by the soil (residue from previous crops) and legumes.

Tables and online calculation assistance are available in the different European countries. For example:
– Belgium: https://protecteau.be/fr/nitrate/agriculteurs/fertilisation-raisonnee/ferti-prairie (designed for use on Firefox)
– France: http://draaf.bretagne.agriculture.gouv.fr/IMG/pdf/GREN_annexe8-1_prairies_09_03_2017_cle874215.pdf
– Ireland: https://www.fertilizer-assoc.ie/p-k-calculator/calculator/
– The Netherlands: www.bemestingsadvies.nl
– UK: https://www2.dardni.gov.uk/gatewayweb/internet/

When to add nitrogen to a grass-legume mix

When sowing a mix on a new meadow: To support the start-up of legumes that do not yet have nodules, fertilise with half the effective nitrogen dose (20 to 25 kg of nitrogen).

For existing meadows: The amount of nitrogen to be supplied is calculated to supplement the nitrogen of microbial origin (refer to previous paragraph).

At the end of winter: Boost meadow fertilisation with mineral nitrogen to maximise spring growth, even in the presence of up to 40% legumes. Legumes (and their bacteria) begin to grow later than grasses. During this time, grass cannot take advantage of nitrogen of bacterial origin. The precise moment to apply N fertiliser can be calculated by the rule of 200°C cumulative degree days. Periods when spreading is not allowed must also be taken into account. The quantity applied depends on the vegetation and the type of soil (30 to 50 nitrogen units).

Calculation of 200°C cumulative degree days.

Record daily minimum and maximum temperatures under shelter.

Find the daily average: (T min + T max) / 2.

Add up daily averages above 0°C.

The optimum moment for the first mineral nitrogen application is a cumulative sum of 200°C × days.

Source: https://www.arvalis-infos.fr/azote-mineral-et-prairies-appliquer-la-regle-des-200-cumules-@/view-14015-arvarticle.html

---

How to sow a grass-legume mix

The temperature requirements of legumes during sowing are higher than for grasses. The optimal sowing period is therefore late summer or spring.

Mixed sowing must be carefully prepared because corrections are very difficult: the combined presence of grasses and legumes greatly reduces the options for chemical weed control.
– Prepare a very fine seedbed on clean, packed soil (croskill rolls).
– Sow at a depth of 1 cm (in Belgium, 1,900 seeds/sqm are recommended).
– Graze/mow the first cut of the new reseed early to prevent weeds from rising to seed and eliminate annual species.

The choice of varieties to mix depends on:
– Soil and climate (fertility, depth, pH, risk of drought or flood)
– Management issues (biodiversity, erosion, landscape aspects)
– Desired longevity for the meadow
– Planned use for the future meadow:
– For mowing, avoid ryegrass and tetraploid clover that dry slowly. Alfalfa, diploid purple clover and white clover are best for this use.
– For grazing, prefer trampling-resistant species such as perennial ryegrass, tall fescue or white clover.

▸▸ Liming

M. Komainda, G. Peratoner, N. Schiebenhöfer and J. Isselstein

Liming is part of the grassland management. It improves the soil fertility, the water holding capacity, the soil structure, the soil biology, the nutrient status and the pH of the soil. Low soil pH levels may be detrimental for the growth of plants, which is mainly due to the release of aluminium, the immobilization of phosphate, the increased leaching of cations (i.e. Mg and Ca) and a reduced microbial activity.

However, under some conditions excess pH levels also induce detrimental effects due to iron deficiency (Figure 4.6) or excess mineralization of organic soils. Limes contain varying amounts of nutrients, of which calcium oxide (CaO) and magnesium oxide (MgO) are the most prominent. Liming is primarily utilized to increase the soil pH and the liming effect of an applied lime is expressed in proportion (%) to the liming effect of CaO. Besides increased pH levels, the CaO improves the soil structure due to its ability to form clay-humus-complexes. Table 4.2 gives an overview on frequently applied limes.

Table 4.2. Forms of applied limes (according to Schilling, 2000 and Schubert, 2006).

Lime	Form	Liming effect (% CaO)	Minor components
Limestone	$CaCO_3$	42-53	Silicate
Limestone with Mg	$CaCO_3$, $MgCO_3$	42-53	Silicate
Burnt lime	CaO	65-95	MgO
Slaked lime	$Ca(OH)_2$	60-70	$Mg(OH)_2$

Due to the negative externalities arising from low pH levels and the disadvantages of excess pH, correct determining the liming demand is of the utmost importance. The possibilities to quantify the demand are shown in Figure 4.5.

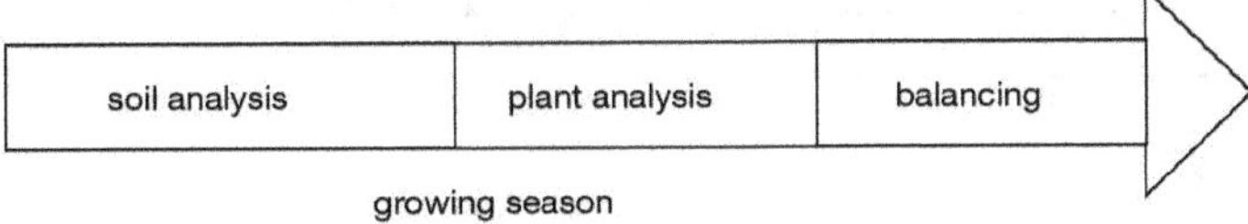

Figure 4.5. Determination of the liming status of respective soils, plants and respective balancing. The analyses should be conducted regularly.

Soil samples are taken at depths between 10 and 30 cm prior to any fertilisation or liming activity to quantify the actual status of the soil regularly (e.g. in spring). The samples should be kept cool until analysis. Figure 4.6 gives an overview of the scheme for soil sampling. Approximately 20 sampling points per hectare are required at minimum.

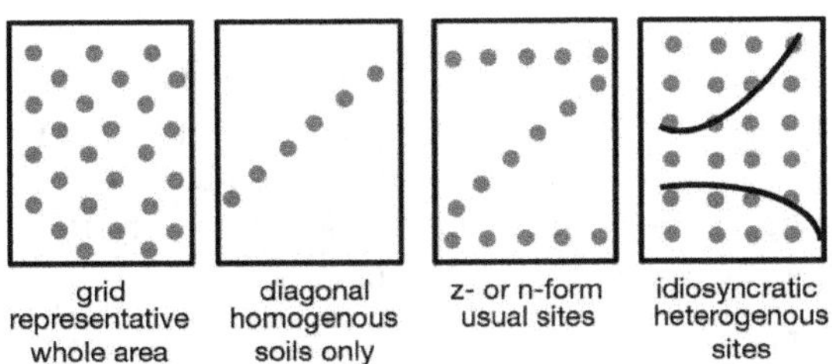

Figure 4.6. Several options are possible for a representative soil sampling on a grassland or arable field site. Derived according to Schubert (2006) and Schilling (2000).

Plant analyses are usually taken at harvest on fresh biomass, in silages or in hay. Every field/pasture should be sampled adequately. Both soil and plant samples should be analysed for CaO content by accredited laboratories. Farm advisors can help find appropriate laboratories.

Liming demand, however, is also a function of the clay and the humus content of the soil. Agricultural soils may contain more than 30% humus, whereas typically values between between 5 and < 15% are found (organic soils are higher). For intensively managed grasslands, approximately 3 to 4 dt CaO/ha are taken up annually by the grass sward and exported via harvested products. A general conclusion on the CaO demand is consequently challenging if no samples are taken. Therefore, precise knowledge on the CaO content (soil and or plant) enables exact calculation of the CaO export. In view of a large share of dairy farms located on sandy soils, Table 4.3 highlights the CaO demand for a soil with a humus content ≤ 15%. The presented amounts are required annually to remain in a good soil status. However, to minimize labor, liming in a three-year-period is possible due to low leaching losses.

Table 4.3. Example of liming requirement of grassland on optimally supplied soil (i.e. optimal pH range). Liming only compensates for the exported CaO required (fertilising guideline for Northern Germany, 2018).

Soil texture	pH range	CaO dt/ha/year
Sand	4.7 - 5.2	4
Loamy-sand	5.4 - 6.0	5
Sandy to silty-loam	5.6 - 6.3	6
Clayey	5.7 - 6.5	8

▸▸ Use of animal manure

M. Komainda, G. Peratoner, N. Schiebenhöfer and J. Isselstein

General considerations

Ruminant livestock producers feed up to 100% grass in the diets and grassland systems for ruminant livestock production are associated with regular export of nutrients within the harvested biomass or animal products. Simultaneously, livestock production is associated with regular and ongoing supply of organic fertilizers, either as manure from the housing or as deposits on pastures. Up to 90% of ingested N is recycled via dung and urinary excretion of grazing animals (Haynes and Williams, 1993). Therefore a close relationship between organic fertilizers and grassland production is found frequently as close nutrient cycle. However, there is a gap between available nutrients and the recycling in plant biomass.

Generally, to meet crop requirements, several nutrients are applied continuously of which nitrogen (N) is the most prominent (see section Main macro-nutrient: N P K). In total 110 Mio. tonnes of mineral N fertilizer are spread world-wide annually on agricultural land, followed by potassium and phosphate (Faostat, 2016). Excess

N applied to agricultural systems may discharge. Initiated by (de)nitrification, emissions of nitrous oxide (N_2O) and nitrogen monoxide (NO) induce climate change and ozone depletion, while volatilization of ammonia (NH_4/NH_3) causes harmful air quality damage, eutrophication or indirectly climate change. During the last 50 years, agricultural greenhouse gas (GHG) emissions increased significantly by 96%, which are linearly related to the N intensity of production. Leaching of nitrate (NO_3) from the root zone and runoff NH_4 contaminate groundwater, rivers, lakes and marine ecosystems, causing harmful externalities for biodiversity and drinking water (Good and Beatty, 2011; Wba, 2016). Derived from national N balances for the EU-27 states, Van Grinsven et al. (2014) calculated an average N use efficiency of only 30% for the agricultural sector. Nitrogen loss abatement by improved plant N uptake and increased N retention is therefore highly required. To reduce and prevent nitrate discharge, the EU nitrates directive (ND, 91/676/EWG) was initiated, prescribing a maximum area based application rate of 170 kg N/ha/year from organic fertilizers or lower. Additionally, phosphorus (P) has become an important negative externality for surrounding ecosystems due to detrimental effects for water (Wageningen UR, 2014). In future the P management will be restricted widely if the accurate utilization on farms will not be adapted soon. A good planning of the fertilization consequently is paramount for high nutrient efficiency (i.e. low losses and high yields).

Organic fertilisers reduce fertiliser bills

In view of the environmental issues, there is a need to use organic fertilisers very efficiently. This is especially important for grasslands as high amounts of nutrients digested by livestock are directly related to the feeding of grass products. High uptake efficiency also reduces mineral fertiliser costs. High N uptake is reached by adapted fertiliser type, amount, timing, application technique and placement (Dinnes *et al.*, 2002; Quakernack *et al.*, 2012). Adequate fertilisation timing emphasises the spring and harvests. Fertilization after the last cut (i.e. in September or October) contrarily increases losses by up to 25% of applied N (Smith *et al.*, 2002).

North-western Europe is favorable for dairy production. Except for Ireland, the proportion of cut grassland dominates feed rations for intensive dairy cow production, with up to six cuts per year. Fertiliser demand is derived from the respective forage yield, as plants accumulate nutrients as function of their biomass dry-matter (DM) production. Overfertilisation increases the losses to the environment and therefore costs. Moreover, it results in a deterioration of the botanical composition with an increase of nitrophilous weeds. Consequently, good knowledge of the yield level is crucial to achieve appropriate fertilisation and to save money. For very good managed grassland plots in experiments, gross yields of up to 16 t DM/ha/year were reported (Nevens and Reheul, 2003; Trott *et al.*, 2004). In agricultural practice, however, yields frequently range between 8 and 10 t DM/ha/year (Van den Pol-van Dasselaar *et al.*, 2015). Grassland soils with good nutrient contents should be supplied with the amounts exported. To supply the sward with nutrients, knowledge on the distribution of the grass yield is important. Table 4.4 gives an overview for several management intensity scenarios.

Table 4.4. Distribution of the DM yield (%-yield share of the first–sixth cut) as function of cutting frequency (www.gruenland-online.de, unpubl. data from the Laimburg Research Centre, adapted).

Cutting frequency (cuts/year)	Contribution of each cut to the annual forage yield (%)					
	1st cut	2nd cut	3rd cut	4th cut	5th cut	6th cut
2	62	38				
3	47	28	26			
4	40	25	20	15		
5	30	20	20	17	13	
6	30	25	15	15	10	5

According to Table 4.4, the first two cuts deliver between 55% and 75% of the total annual yield. Appropriate fertilisation practice harmonises crop demands and nutrient delivery for each harvest.

Planning of organic fertilisation

The primary nutrients delivered in organic fertilisers are N, phosphorus (P_2O_5 or P), potassium (K_2O or K) and magnesium (MgO), while CaO, sulphur, copper, manganese, zinc, boron and molybdenum may also be available. Long-term dose-response trials derived reliable correlations between yield and nutrient demand. Approximately, the demand for intensively managed grassland is approximately 2.5 kg N, 0.8 kg P_2O_5, 3.1 kg K_2O and 0.45 kg MgO per 100 kg DM yield in a four-cut system. Consequently, the area-based nutrient demand increases with the yield level as shown in Table 4.5. The nutrient demand shown in the table only applies to intensive production. The fertilisation of intensive grassland aims to meet the demand to ensure high yields and quality. It should not be used to compute the demand of species-rich grasslands which do not need any external nutrient inputs. If nature conservation (i.e. high biodiversity) is focused on grasslands, the nutrient input via fertiliser should be avoided; otherwise, species will disappear from the swards.

Table 4.5. Nutrient demand of grassland of varying yield levels. The mineralization is taken into account.

Yield level [t DM/ha]	N [kg/ha]	P_2O_5 [kg/ha]	K_2O [kg/ha]	MgO [kg/ha]
6	150	48	186	27
7	175	56	217	31
8	200	64	248	36
9	225	72	279	41
10	250	80	310	45

Fertilisation strategy

According to the Nitrates Directive (1991), a maximum supply of 170 kg N/ha/year derived via organic fertilisers may be applied on farms. In groundwater sensitive regions, i.e. nitrate vulnerable zones (DEFRA, 2018), this value may be lower. Advisors can be consulted for local values. The nutrient contents of organic fertilisers are highly variable based on management, feeding, year and time of the year. An important prerequisite for accurate organic fertilisation is the chemical analysis of the available fertilisers on the farm. Table 4.6 shows some average values for specific organic fertilisers as a rough estimate. These values, however, do not apply to specific farming situations.

Table 4.6. Nutrient contents of some organic fertilisers (derived from: Baumgartner *et al.*, 2006; Elsäßer, 2009; Klocker *et al.*, 2017; Agricultural chamber of North Rhine-Westphalia, 2018). These values give only rough estimates. The actual contents of organic fertilisers on each farm may vary considerably depending on management, feeding, year and time of the year. Annual analyses of fertiliser samples are required for optimal fertilisation.

Organic fertiliser	Total N	NH$_4$-N	P$_2$O$_5$	K$_2$O	MgO
	kg/m³ or kg/t	kg/m³ or kg/t	kg/m³ or kg/t	kg/m³ or kg/t	kg/m³ or kg/t
General fertiliser					
Liquid manure	3.5	1	1.9	5.8	1.1
Solid manure	4.2	0.3	3.5	6.1	2.2
Slurry	2.7	1.7	0.1	9	0.4
Biogas residue	3.2	1.5	1.8	5.8	1
Liquid manure					
Cattle					
Young stock, grass-based	3	-	1.2	4.7	-
Young stock, arable-based	2.4	-	1	4	-
Dairy cows, grass-based	3.7	-	1.4	5.3	-
Dairy cows, arable-based	3	-	1.3	4.3	-
Beef cattle	3.6	-	1.5	3.7	-
Solid manure					
Cattle	3.2	-	2.9	5.9	-
Horse	2.3	-	1.5	3	-
Sheep	4.3	-	2.1	4.9	-
Pigs	5.8	-	5.1	5.5	-

Generally, the fertilisation level follows actual demand, i.e. yield. For some nutrients, however, excess fertilisation one year to increase the soil storage may be valuable (i.e. lime). The nutrient requirements are determined by the yield, and the yield level should be known for each site and harvest. Additionally, it is important to understand that only parts of the nutrients in organic fertilisers will become available for plants during the year of application. For liquid organic dairy manure, for instance, approximately 50% of the total N content or at least the NH_4-N share will become available in the year of application. Approximately 10% of the total slurry-N may become available during the subsequent year. One example of a grassland N-fertilisation strategy is given as follows:

N fertilisation (kg N/ha) = N demand (kg/ha) – soil N (kg/ha) – N from legumes (kg/ha) – N from organic fertiliser N preceding year (e.g. 10% of organic fertiliser N of the preceding year)

which means, according to Table 4.5 and a yield level of 10 t DM/ha:

N fertilisation = 250 - 10 - 15 - 17 = 208 kg N/ha

A value of 17 kg N/ha from organic fertiliser of the preceding year means that 170 kg N/ha were applied in that year (i.e. 10% of preceding total organic N fertilisation).

Here the soil N refers to the expected mineralisation during the growing season, which considers the soil humus content in the system in Germany. The proportion of legume-N is higher under less intensive management and in regrowth after the first harvest. In relation to the time of the growing season, the climatic and soil conditions, and the leguminous species, between 3 to 6 kg of N are supplied per %-share to the yield. The N fertilisation scheme varies in relation to country-specific regulations. The example refers to regulations formulated in Germany. For grazing systems in Ireland the additional N input is based on the stocking rate in relation to the timepoint of the growing season (Teagasc, 2018). In Sweden the Advisory Board for Agriculture is responsible for oversight. Ignoring good fertilisation practice guidance may be penalised by local authorities, so appropriate consideration of regulations regarding fertilisation practices is essential. Advisors can help ensure regulations are properly met.

Table 4.7. Fertilisation grassland at a yield level of 10 t DM/ha with liquid dairy manure and resulting mineral external fertilisers. Example 1 refers to above mentioned calculation of N demand.

Example of nutrient content of liquid dairy manure		Specific nutrient demand	Nutrient demand	N demand expl. 1	Possible manure rate	Nutrients applied with 50 m³/ ha manure	Mineral fertiliser
Nutrient	kg/m³	kg/100 kg	kg/ha	kg/ha	m³/ha	kg/ha	kg/ha
Total-N	4	2.5	250	208			
NH_4-N	2.3				90	115	93
P_2O_5	1.5	0.8	80	80	53	75	5
K_2O	4.5	3.1	310	310	69	225	85
MgO	1	0.5	50	**50**	**50**	50	0

Table 4.7 provides an example for the application of fertilisers to grassland with a yield level of 10 t DM/ha. In this example a liquid dairy manure with the listed contents of

nutrients is applied. The plant demand refers to a four-cut grassland for conditions in Germany. Again, in other countries the specific crop requirement may be lower or higher. According to Table 4.7, in total 50 m³/ha/year of slurry are applied due to the demand of MgO which is met with this amount. If the total grass demand exceeds the amounts of nutrients supplied with organic fertilisers, additional mineral fertilisers may be used for the respective cuts. For the share of each cut to the total annual yield, refer to Table 4.4.

Excessive potassium fertilisation may have detrimental effects on livestock. A threshold for maximum fertilisation is 100 to 120 kg N/ha and 150 kg K_2O/ha. The recovery of N further depends on the application technique with shallow injection or acidification increasing the N contribution by up to 20% in comparison to broad-cast application. For further information about the adequate handling of fertilisers, refer to advisory tables, booklets, statements and recommendations (e.g. agricultural chambers, DEFRA, Teagasc, Jordbruksverket, country ministries etc.).

▸▸ Other macro- and micronutrients

M. Komainda, G. Peratoner, N. Schiebenhöfer and J. Isselstein

On grazed grasslands the nutrient supply via grazing animals should be taken into consideration when planning fertilisation. On grazed land the demand of P, K, Mg and sulphur (S) is usually much lower than on cut grassland.

S represents an important nutrient for grasslands. S depletion mainly occurs when above-average precipitation during winter causes S leaching. In spring the analysis of soil samples for the content of mineralised S (S_{min}) is a good indicator of the fertilisation requirement. For efficient S fertilisation, however, an adequate and representative forage sample should be analysed. Inadequate S supply is indicated by a nitrogen-to-sulphur (N:S)-ratio of $\geq$ 15:1. According to this, S fertilisation may be required. Sulphur deficiencies can be expected at high supply levels of mineral N, while they are less likely if organic fertilisers are used. Potassium, magnesium and natrium affect each other in the plant uptake and animal digestion. Generally, 1% potassium in DM is sufficient to meet livestock requirements. Most soils, however, are low in potassium content due to leaching, so fertilisation is required in managed grasslands. However, if plants have accumulated excessive potassium, it may cause metabolic disorder in animals due to antagonism with magnesium and consequently lead to grass tetany. The incidence of grass tetany increases with magnesium contents < 0.2% in DM and at > 3% potassium and > 20% crude protein in DM. This can occur in very young plant material early in the year. Clay soils are usually rich in potassium and do not require any external input. Acidic and light-sandy soils are usually low in magnesium, whereas volcanic and clay soils have higher contents. Soils rich in natrium are found in the heavy marsh lands located near the coastal line as well as in soils in aridic zones. Potassium and natrium may be antagonistic. Natrium is important for livestock and should be supplied separately because it may block physiologic processes in plants, which are induced by potassium. Selenium (Se) is also important for livestock. For lactating dairy cows, concentrate usually contains selenium. Selenium content in plants mainly depend on the selenium content of the soil. In areas with low selenium soil content, the supply for the animals cannot be ensured via grass biomass without external supplementation. For extensively reared beef cattle or horses, selenium deficiency is frequent. Animals require between 0.1 and 0.3 mg Se kg/DM. Due to toxic effects, maximum selenium fertilisation is limited to 8 g/ha. For instance, Klotz *et al.* (2012) demonstrated how to fertilise selenium by mixing with liquid manure, which reduces application costs. Selenium fertilisation is difficult on soils with a low pH. As a result, adequate liming is important. Micronutrients should be applied if a deficiency is recognised after plant or soil analysis. Predominantly forage plants grown on arable land, such as lucerne, may have demand for specific micronutrients (e.g. molybdenum). Advisors can help choose appropriate fertiliser products. Figure 4.7 gives an overview of the relationship between plant availability and micronutrients. Table 4.8 gives an overview for authorities on fertilisation law in some EU Member States.

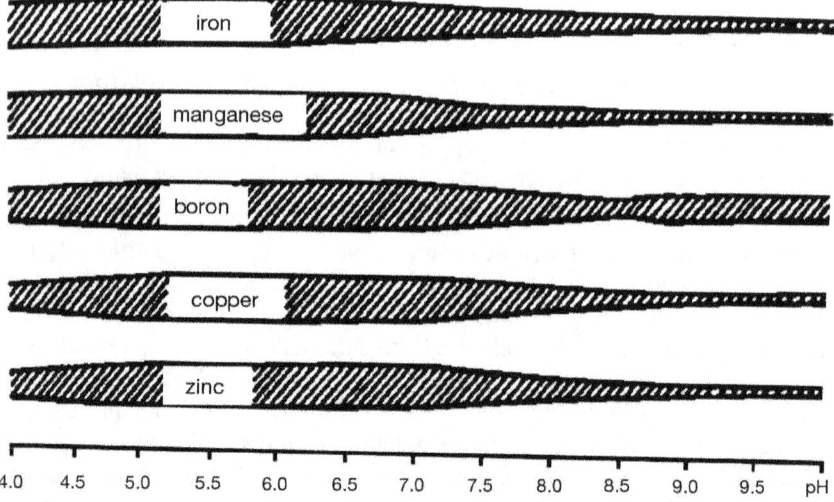

Figure 4.7. Plant availability of micro nutrients in relation to the pH of the soil. The width of each strip represents the solubility (adapted from Schilling, 2000).

Table 4.8. Overview of responsible agricultural authorities in each Inno4Grass partner country (date: 19.12.2018).

Country	Responsible for agricultural issues	Webpage
Belgium	Foreign Affairs, Foreign Trade and Development Cooperation	https://diplomatie.belgium.be/en/policy/coordination_european_affairs/policy/agriculture_and_fisheries
France	Ministry of Agriculture	http://agriculture.gouv.fr/english-contents
Germany	Landwirtschaftskammer, Landesamt für Landwirtschaft, Ministry of Agriculture	https://www.lwk-niedersachsen.de/, https://www.bmel.de/EN/Homepage/homepage_node.html
Ireland	Teagasc	https://www.teagasc.ie/contact/
Italy	Ministry of Agriculture	https://www.politicheagricole.it/flex/cm/pages/ServeBLOB.php/L/IT/IDPagina/202
Netherlands	Ministry of Agriculture, Nature and Food Quality	https://www.government.nl/ministries/ministry-of-agriculture-nature-and-food-quality
Poland	Ministry of Agriculture	https://www.gov.pl/web/rolnictwo/
Sweden	Swedish Board of Agriculture	https://www.jordbruksverket.se/swedishboardofagriculture/engelskasidor/crops/plantnutrients.4.6621c2fb1231eb917e680003205.html
List for responsible authorities in EU Member States		https://ec.europa.eu/agriculture/links-to-ministries_en

Chapter 5

Environment and biodiversity

*Agnes van den Pol-van Dasselaar, Felicitas Kaemena
and Leanne Bastiaansen-Aantjes*

▶▶ Carbon sequestration

A. van den Pol-van Dasselaar and F. Kaemena

Carbon sequestration as an ecosystem service

Grasslands are well-known for their contribution to food production (milk, meat). Food production is an important service that grasslands deliver. Grasslands deliver many other services and goods as well, which are called ecosystem services. Ecosystem services are the benefits that humankind gains from its interaction with natural resources.

Ecosystem services from grasslands can be divided in four types (MEA, 2005):
– Provisioning services: products obtained from ecosystems, e.g. production of food, water
– Regulating services: benefits obtained from the regulation of ecosystem processes, e.g. climate and disease control
– Cultural services: non-material benefits people obtain from ecosystems through spiritual enrichment, cognitive development, reflection, recreation, and aesthetic experiences, e.g. recreation and beauty of the landscape
– Supporting services: ecosystem services that are necessary for the production of all other ecosystem services, e.g. nutrient cycles, crop pollination

Ecosystem services are important for farmers and for society in general. Some examples of ecosystem services that grasslands deliver are shown in Figure 5.1. Carbon sequestration is one of them.

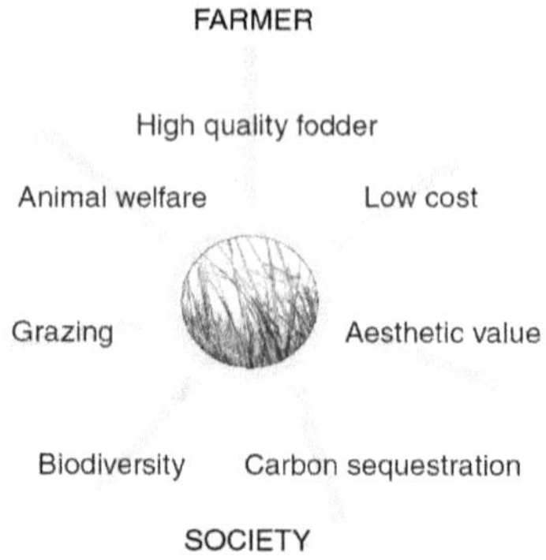

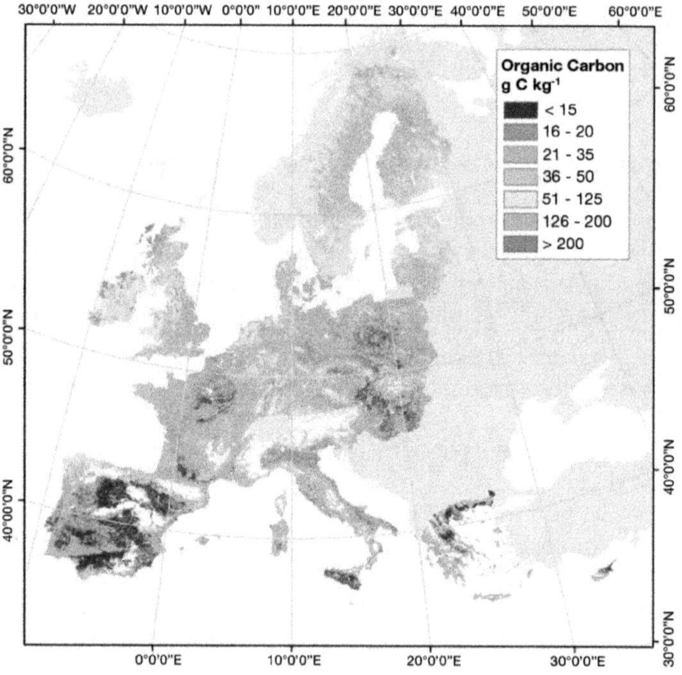

Figure 5.1. Selection of ecosystem services that grasslands provide to farmer and society (Van den Pol-van Dasselaar *et al.*, 2018a).

Carbon storage

Soils are important to combat climate change since they are capable of storing enormous amounts of carbon (C). They act as a huge C reservoir. Figure 5.2 (based on modelling) shows the differences in soil organic C content between northern and southern Europe. The organic C content of soils is higher in the north of Europe than in the south. This is mainly due to abiotic factors (climate, etc.).

Figure 5.2. Map of predicted topsoil organic carbon (C) content (g C/kg) (De Brogniez *et al.*, 2015).

Along with abiotic factors, management factors also play a role in C storage and C sequestration. The C storage capacity of grasslands is much higher than that of arable lands. If grasslands are ploughed, considerable amounts of C are lost. Accordingly, to combat climate change, it is important to maintain current C stocks and prevent ploughing of grasslands as much as possible.

Effect of grassland management on carbon sequestration

Many types of grasslands have not yet reached their maximum storage capacity and are able to store additional C. This additional C storage is called carbon sequestration and is crucial in combatting climate change. But, as mentioned previously, even when no additional C is sequestered, grasslands are very important with regard to climate change because they store enormous amounts of C. The extent to which additional C can be taken out of the atmosphere by grasslands and stored in the soil will determine the overall role of grasslands in mitigating the impact of increased emissions.

A literature review of the EIP-AGRI 'Grazing for carbon' Focus Group (Van den Pol-van Dasselaar *et al.*, 2018b) showed that there is net C sequestration within grassland systems in general, but in a mixed grazing and cutting system there is less C sequestration than under a pure grazing system (Figure 5.3).

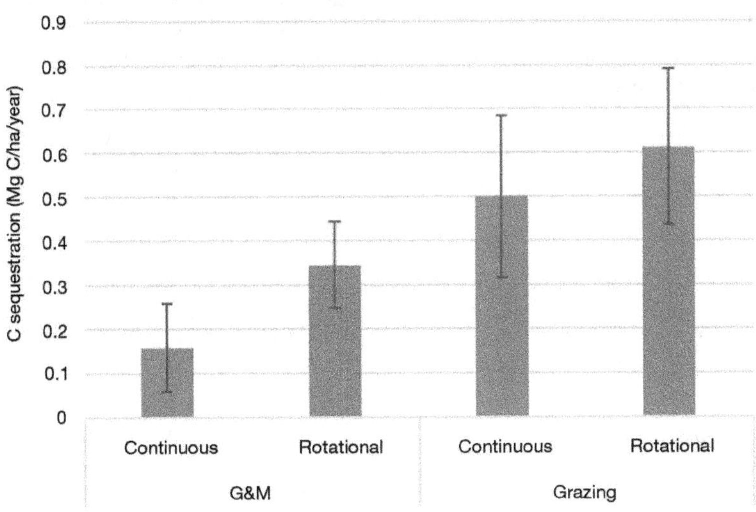

Figure 5.3. Mean carbon (C) sequestration rate (Mg C/ha/year) for mixed grazing and cutting systems (G&M) or grazing-only systems (Grazing) (results of literature review Klumpp *et al.* in Van den Pol-van Dasselaar *et al.*, 2018b).

Soil management and abiotic factors both affect C equilibrium and C stock. The key challenge for sustainable grazing livestock systems is to identify the optimal type of management to combine animal production with the delivery of other ecosystem services like C sequestration.

The effect of grazing on C sequestration is rather complex and is affected by grazing intensity (Figure 5.4). Many processes play a role, and these are each individually affected by abiotic factors and management factors. The effects of grazing are driven by plant tissue removal (defoliation), excretion (urine and dung deposits) and trampling, which exerts mechanical pressure and causes physical damage to the vegetation where animals pass repeatedly. In the short term, grazing results in a reduction in aboveground standing biomass, as well as changes in plant nutrient status. If there is considerable dead plant material in the sward that is shading the live leaves, grazing can allow light to penetrate into the plant canopy and encourage new tiller formation, enhancing primary productivity. Conversely, if grazing is too intense or the period between successive grazing events is too short, the amount of live leaves can be reduced so much that light interception falls, growth/carbon capture is reduced and litter production is low (i.e. reduction in C inputs to soil). Between these two extremes, there is relatively little change in growth with changes in grazing pressure. However, the quality of the herbage and the production of litter still respond to changes in grazing pressure within this range. Higher grazing pressure increases pasture regeneration and herbage quality (as long as there is sufficient N available), but reduces litter production, and vice versa. There is a trade-off between quality (promoting animal production) and litter production (promoting C sequestration). What constitutes low/medium/high grazing pressure varies between locations and over time; the lower the pasture growth, the lower the grazing pressure or the longer the period between grazing events, and vice versa. The key aim for sustainable grazing livestock systems is to determine the optimum stocking rate where the optimum grass intake coincides with a certain amount of C sequestration in the soil.

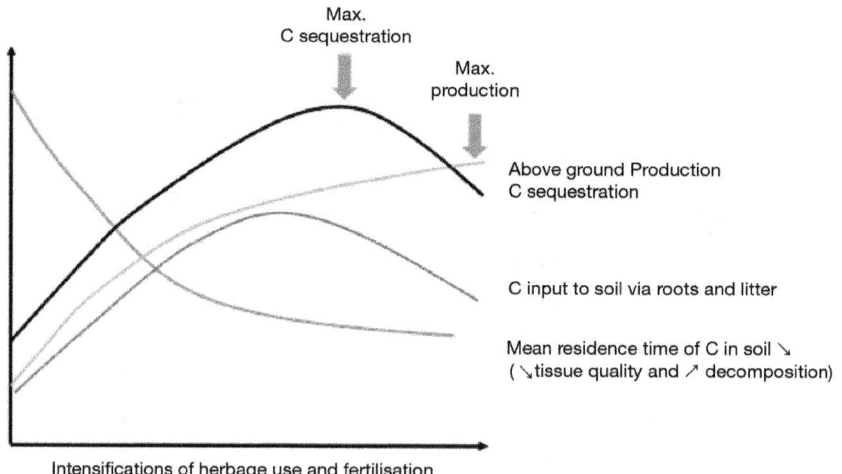

Figure 5.4. Effects of grassland intensification by grazing, cutting and fertilisation on C inputs, mean residence time of soil organic C and C sequestration (adapted from Soussana and Lemaire, 2014).

▸▸ Greenhouse gas emissions

A. van den Pol-van Dasselaar and F. Kaemena

Climate change

Over the last decades, increasing attention has been paid to the effects of climate change. Moderate warming and more carbon dioxide in the atmosphere may help some plants to grow faster. However, more severe warming, floods and drought may reduce yields. Livestock may be at risk, both directly from heat stress and indirectly from reduced quality of their food supply.

To deal with climate change in agriculture, two pathways are possible:
− Mitigation
− Adaptation

Mitigation options in agriculture are options which reduce the emissions of greenhouse gases (CO_2, N_2O and CH_4) from agricultural production systems.

Adaptation options describe ways for agricultural production systems to adapt to future climate conditions (like global warming, larger climatic variability and increased frequency and severity of droughts and floods). Often mitigation options and adaptation options interact.

Options to mitigate greenhouse gas (GHG) emissions from animal production systems are strongly linked to the N and C cycles in those systems. The four key components with their respective main GHGs are:

• Manure/fertiliser: mainly N_2O and CH_4 emissions;

• Soil: mainly CO_2 and N_2O emissions;

• Crop/feed: mainly N_2O emissions;

• Animal: mainly CH_4 emissions (as a result of enteric fermentation).

The sources and sinks of the various GHG emissions from animal production systems have been identified and the variation in their size has been evaluated. Many mitigation options have been tested experimentally and the results have been documented in several reviews (e.g. Vergé *et al.*, 2007). Models have been developed to predict GHG emissions and to evaluate mitigation strategies. In a similar manner adaptation options have been studied (e.g. Olesen *et al.*, 2011), which is particularly important for areas which are most vulnerable to climate change. Research results show numerous interactions between mitigation and adaptation in the context of different environmental and socio-economic conditions. Generally, limited information is available on the quantification and comparison of synergies and trade-offs, and few papers report on this (e.g. Smith and Olesen, 2010).

Mitigation and adaptation options

Van den Pol-van Dasselaar and Bannink (2014) provided a qualitative overview of mitigation and adaptation options in livestock production systems, and of their

synergies and the trade-offs between individual GHGs (Table 5.1). It is based on a review of available literature and expert judgement. The options are strongly linked to changes in the N and C cycles of the farming system. Four categories of options are distinguished in Table 5.1, at the level of:
– manure/fertiliser;
– soil;
– crop/feed;
– animal.

Many adaptation and mitigation options in Table 5.1 are linked to grasslands. Since synergies and trade-offs between GHG exist for adaptation and mitigation options, accurate predictions of the effects of these options are needed to tailor them in the context of specific farming conditions. Climate change may cause reduced efficacy or applicability of mitigation strategies, and may lead to lower yields due to elevated temperatures and fluctuations in water availability. Furthermore, in many countries, the impact of agriculture on climate is a less important issue due to socio-economic reasons, such as when addressing famine (Vergé *et al.*, 2007).

Because of the large contribution of N_2O in agricultural GHG emissions, it is clear that the botanical composition of swards (especially the proportion of legumes), the management of manure and fertilisation, and the share of concentrates will play key roles.

Table 5.1 focuses on options at the field and the animal scale. It is important to have a clear understanding of the possible options at that scale, since it is the scale where farmers make their day-to-day decisions. However, it is also important not to forget the regional and global effects, since decisions at the scale of field and animal will affect the global scale as well. For example, the impact of rising food demands means, other things being equal, that a reduction in food production in a certain region would result in increased food production elsewhere. This can result in a net increase in global GHG emissions, if the countries expanding food production were unable to produce food with low emission intensity (Schulte *et al.*, 2011).

Table 5.1. Qualitative overview of mitigation and adaption options in livestock systems at the level of manure/fertiliser, soil, feed/crop and animal, and their synergies and trade-offs (Van den Pol-van Dasselaar and Bannink, 2014).

Option	Effects on mitigation (+)	Effects on adaptation (+)	Mit.pot. CH_4 (-/+/++)	Mit.pot. N_2O (-/+/++)	Mit.pot. CO_2 (-/+/++)	Mitigation variability (variable/robust)	Importance in mitigation (low/high)	Adaptation variability (variable/robust)	Importance in adaptation (low/high)	Productivity impacts (-/+)	Costs (high/low/?/benefit)	Future measure / ready to use	Applicability by farmers (easy/difficult)	Acceptability for farmers (poor/good)
MANURE / FERTILISER														
Fertilisation rate	+	+	−	++		rob	high	rob	low	+	ben	ready	dif	good
Fertiliser type	+			+		var	high			+	ben	ready	dif	good
Fertiliser application	+			+		rob	high			+	ben	ready	dif	good
Cover slurry stores/manure heaps	+		++	−	+	rob	low				low	ready	easy	good
Manure cooling	+		+			rob	low				high	future	dif	poor
Manure treatment	+		++	−	+	rob	high				?	ready	dif	good
Filtering CH4 from barns	+		+	+		var	low				high	future	dif	poor
SOIL														
Reduced/zero-tillage	+	+		−	++	rob	high	rob	high		low	ready	dif	poor
Prevent soil compaction	+			+	+	var	low			+	low	ready	dif	good
Water management	+	+		+	+	var	low	var	low	+	low	ready	dif	good
Irrigation		+		+		var	low	var	low	+	low	ready	dif	good
Restoring degraded lands	+	+		+	++	rob	high	rob	high	+	high	ready	dif	poor
Pasture reclaiming/recovery	+				++	rob	high				low	ready	easy	good
Incorporation crop residues		+		−	+	rob	low	rob	low	+	?	ready	easy	poor
CROP / FEED														
Crop rotation	+	+	+	+	++	rob	low	var	low		low	ready	easy	good

Option	Acceptability for farmers (poor/good)	Applicability by farmers (easy/difficult)	Future measure / ready to use	Costs (high/low/?/benefit)	Productivity impacts (-/+)	Importance in adaptation (low/high)	Adaptation variability (variable/robust)	Importance in mitigation (low/high)	Mitigation variability (variable/robust)	Mit.pot. CO_2 (-/+/++)	Mit.pot. N_2O (-/+/++)	Mit.pot. CH_4 (-/+/++)	Effects on adaptation (+)	Effects on mitigation (+)
Perennial crops	good	easy	ready	low				low	rob	+	+			+
Legumes and mixtures	good	easy	ready	ben	+			low	rob	+	++	−		+
New pasture species	good	dif	future	low	+			low	var			+		+
Improved crop varieties	good	dif	future	low	+			low	var		+	+		+
Novel crops	poor	dif	future	low		low	var	low	var			+	+	+
Cover crops	poor	easy	ready	low	+	low	rob	low	rob	++	+		+	+
Conversion to grass	good	easy	ready	low				low	rob	++				+
Reforestation	poor	dif	ready	high	−			low	var	+				+
Optimal forage management	good	easy	ready	ben	+			low	var	+		+		+
Biodiversity	poor	easy	ready	low	+	high	var						+	
Plant breeding	good	easy	future	high	+	high	rob						+	
Use climate forecasting	good	easy	ready	ben	+	high	var						+	
Different planting dates	good	easy	ready	low	+	low	var						+	
Conservation as a buffer	good	easy	ready	low		low	rob						+	
Mixed versus single species grass	good	easy	ready	low	+	low	var						+	
Agroforestry	poor	dif	ready	high		low	rob	high	rob	++			+	
Optimal grazing	good	dif	ready	ben	+	high	var	low	var	+			+	+
Increased feed digestibility	good	dif	ready	high	+			high	rob			++		+
Feed analysis	good	easy	ready	high				low	rob		+	+		+
Improving roughage quality	good	dif	ready	low	+			high	rob			+		+

Option	Acceptability for farmers (poor/good)	Applicability by farmers (easy/difficult)	Future measure / ready to use	Costs (high/low/?/benefit)	Productivity impacts (-/+)	Importance in adaptation (low/high)	Adaptation variability (variable/robust)
More concentrates	good	easy	ready	high	+		rob
Improving grass quality	good	dif	ready	low	+		rob
Use of silage maize	good	easy	ready	low			rob
Additives in general	poor	easy	future	high			var
Additive nitrate	poor	easy	ready	high			rob
Matching supply and demand	good	dif	ready	ben	+		rob
Supplemental feeding	good	easy	ready	high	+	high	rob
ANIMAL							
Rumen control via breeding	poor	dif	future	high			var
Immunological control	poor	dif	future	high			var
Less consumption animal products	poor	easy	ready				rob
Increased production in general	good	dif	future	ben	+		rob
Incr prod extensive systems	good	easy	ready	ben	+		rob
Incr prod intensive systems	good	dif	future	ben	+		rob
Animal breeding	good	dif	future	high		high	var
Animal management	good	dif	ready	ben			var
Animal manipulation	poor	easy	ready	high			var
Replacement rate cattle	good	dif	future	ben	+		rob
Cooling of animals	good	easy	ready	high		high	rob
Livestock mobility	good	dif	ready	low		high	var
Animal health	good	dif	future	low	+	low	var

Option	Importance in mitigation (low/high)	Mitigation variability (variable/robust)	Mit.pot. CO_2 (-/+/++)	Mit.pot. N_2O (-/+/++)	Mit.pot. CH_4 (-/+/++)	Effects on adaptation (+)	Effects on mitigation (+)
More concentrates	high	rob	−	+	+		+
Improving grass quality	high	rob		+	+		+
Use of silage maize	low	rob	−		++		+
Additives in general	low	var					+
Additive nitrate	low	rob			++		+
Matching supply and demand	high	rob		+	+		+
Supplemental feeding	low	var		+	+	+	+
ANIMAL							
Rumen control via breeding	low	var		+	+		+
Immunological control	low	var		+	+		+
Less consumption animal products	low	rob		+	++		+
Increased production in general	high	rob		+	++	+	+
Incr prod extensive systems	high	rob		−	++	+	+
Incr prod intensive systems	high	rob		+	+		+
Animal breeding	low	var		+	+	+	+
Animal management	low	var			+	+	+
Animal manipulation	low	var		+	+		+
Replacement rate cattle	high	rob		++	++		+
Cooling of animals						+	
Livestock mobility						+	
Animal health				+	+	+	

Effect of grazing versus no grazing on GHG emissions

Looking at the potential effect of grazing versus no grazing in relation to climate change, the most important difference between grazing and keeping cows indoors all year is where the dung and urine land: some in the pasture, or all in the cowshed. When dung and urine are deposited in the field, a large amount is deposited on a small area where the minerals cannot be used – at least, not in the short term – and thus losses are more likely. Dung and urine collected from the cowshed can be used as fertiliser. This improves the nutrient use efficiency and reduces the need to buy fertiliser, while yields remain the same. Keeping cows indoors all year can reduce a farm's nitrogen imports by about 50 kg/ha/year compared to grazing. In addition, grazing affects the type of nitrogen loss. During grazing, relatively large amounts of nitrate may be leached and there may be considerable denitrification. Furthermore, there may be relatively large emissions of nitrous oxide (N_2O). In contrast, collecting dung and urine from the stable and spreading it on the land, as is the case when keeping cows indoors all year, results in more ammonia volatilisation. This ammonia volatilisation may be partly reduced by adapting the feed strategy (less protein in the ration). When keeping cows indoors all year, the energy use and hence the CO_2 emissions may also be larger because there is much more use of machinery. The grazing system does not affect methane emissions from grasslands themselves. The larger amount of manure in the slurry pits when keeping cows indoors all year, however, may lead to more methane emissions. The overall effect of grazing is illustrated in Figure 5.5 that shows the greenhouse gas emissions of 46 farms with different grazing intensities (full-day grazing, half-day grazing and access to pasture) (Lasar, 2017). Consideration of the groups and different greenhouse gas sources provides good insight into the relationship between milk production and CO_2 emissions. It becomes clear which areas are particularly sensitive and show a risk of high emissions. Differences between grazing systems are summarised in Table 5.2.

Table 5.2. The effect of grazing on various environmental aspects. The score ranges from - - to + +, with + + signifying that the relevant system scores negative for the point in question, e.g. high leaching or high losses (Van den Pol-van Dasselaar *et al.*, 2008).

Environmental aspect	Unrestricted grazing	Restricted grazing	No grazing
Nitrate leaching, N_2O emission, N losses	+	-	- -
P losses	+	+/-	-
Ammonia volatilisation	- -	-	+/-
Energy use, CH_4 emission	-	+	+ +

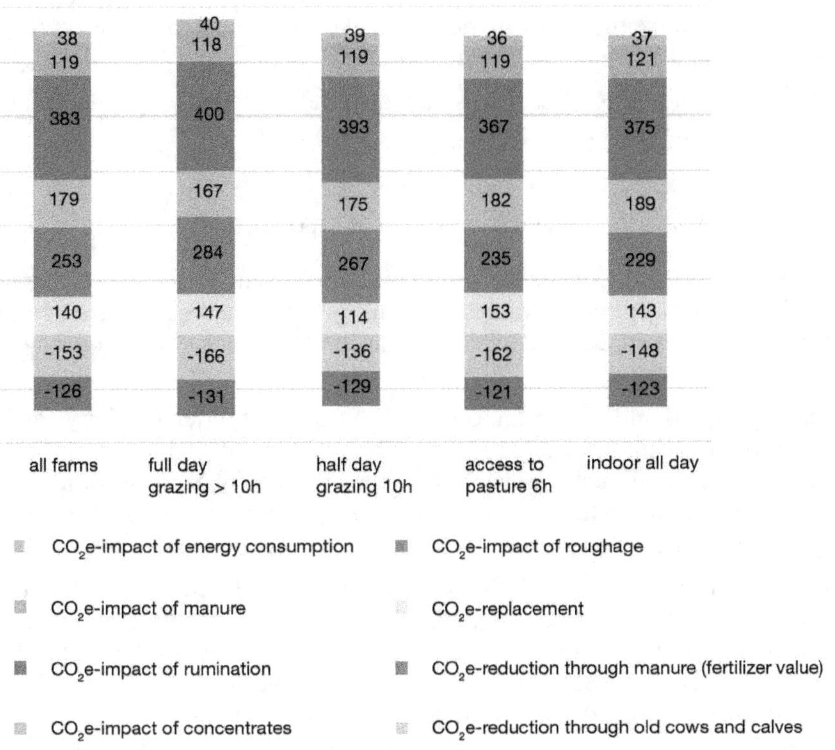

Figure 5.5. Greenhouse gas sources (g CO_2e/kg ECM: CO_2-equivalent is a unit of measurement to compare the effect of different greenhouse gases on warming with that of CO_2) of 46 farms with different grazing intensities: full-day grazing (> 10 hours/day); half-day grazing (> 6 hours/day); access to pasture (< 6 hours/day) and indoor all year.

▸▸ Water quality

F. Kaemena and L. Bastiaansen-Aantjes

Water law requirements: legal background

The EU Water Framework Directive (WFD) forms the regulatory framework for the protection of water bodies in the European Union. One of its aims is to maintain or gradually improve the material and quantitative status of surface waters and groundwater across the board. Special regulations are laid down in the Groundwater Directive as a daughter directive to the WFD. According to this directive, the following quality standards apply throughout Europe for the assessment of good chemical status: 0.1 g/l for active ingredients in pesticides for individual substances and 0.5 g/l as the sum value of all proven pesticides and their degradation products.

Furthermore, the guideline for nitrates specifies a quality standard of 50 mg NO_3/l. If the nitrate limit value of 50 mg NO_3/l is exceeded or 75% of the limit value is reached, measures to reverse the trend are necessary. The limit value of 50 mg NO_3/l groundwater is also laid down in the Nitrates Directive.

According to the Water Resources Act, harmful or detrimental changes to water bodies are to be avoided.

Permanent grassland: groundwater protection

As a general rule, grasslands are found on less efficient soils and are characterised by reliably low N minimum values when using cuttings. As a measure to preserve grasslands, the energetic use of the vegetation to ensure low N output and avoid upheavals is desirable.

Permanent grasslands are primarily used to provide fodder for dairy cattle. They are also of great importance as a landscape-determining element and for resource protection. An additional use may include the provision of biomass for fermentation, especially in regions where other forms of use no longer exist. In general, the use of grasslands for energy purposes is positive if it ensures grasslands are maintained.

The composition of the population varies from grassland to grassland and depends on cultivation (use, fertilisation, maintenance) and location factors (climate, soil, terrain). Experience from water protection consulting shows that N surpluses decrease with increasing cutting use. Maintenance measures should ensure the long-term maintenance of a dense grassland sward so that nutrient yields remain low and a change to grassland renewal can be avoided in the long term. Reseeding serves to eliminate stand gaps and to maintain or establish the desired composition of the grassland sward. It is carried out in March to early April or towards the end of August. The late summer date benefits from the lower growth rate of the old sward, but requires sufficient soil moisture. If a renewal of the sward is unavoidable, an uninterrupted procedure with herbicide application and subsequent sowing using the through-seeding technique is recommended. This can largely prevent nitrate leaching as a result of the otherwise unavoidable humus mineralisation and a disturbance of the soil structure. Particular importance must be attached to grassland care.

Low autumn N min values of ~30 kg N/ha and N concentrations in leachate tending towards zero can be measured under grassland. The autumn N min values can be reduced compared to grazing by cutting. From a water protection point of view, grassland conservation is particularly important on sites with high mineralisation.

Conversion of arable land into grassland and extensive grassland management

From the point of view of groundwater protection, permanent grasslands offer several advantages over arable land use:
— Year-round greening with high N uptake

– No tillage, i.e. low rates of mineralisation (exception: reseeding)
– Less use of crop protection agents

Added to this is the very low susceptibility of grassland to erosion. The nitrate emissions under grassland are therefore usually significantly lower than with arable land use. Exceptions are grazing livestock with high stocking densities and grassland upheaval for grassland renewal or transfer to arable land use. In water catchment areas, in addition to the conversion of arable land into grassland, the conservation and extensive management of grassland is also promoted.

Example of Lower Saxony: use of nature conservation, compensation and replacement measures and networking with drinking water protection measures

The success of groundwater protection measures under the Lower Saxony cooperation model must be measured against a large number of target criteria, the most important of which are listed below in key points:
– Maintaining good groundwater quality or, where necessary, improving or rehabilitating groundwater quality by reversing the trend in quality development;
– Achieving good acceptance among the main target groups of land users and the authorities and institutions involved;
– Compliance with the financial framework set by the revenue from the water abstraction fee;
– Permanent establishment of the measures introduced to achieve sustainable groundwater protection.

The Lower Saxony cooperation model has created essential preconditions for the implementation of the abovementioned objectives. In many cases, however, efficient implementation of groundwater protection measures cannot be achieved solely through cooperation with agriculture and forestry, but requires the inclusion of far-reaching nature conservation measures.

Reactivation of wetlands

Lowland soils originally close to groundwater often have large natural nitrogen reserves in the form of fen peat or humus-rich mineral soil horizons. Groundwater subsidence due to drainage measures or groundwater abstraction accelerates the mineralisation of the organic substance and thus releases nitrogen. This process can be accelerated by agricultural land use. For this reason, nitrate concentrations of up to 300 mg NO_3/l are measured in the leachate of fen soils or at the groundwater surface. The main concern of groundwater protection-oriented reactivation of wetlands is the prevention of such 'nitrate breakthroughs' into groundwater. The measures for this concern the control of the groundwater balance and agricultural land use:
– In order to ensure low nitrate emissions from low moor soils, the groundwater level must ensure at least one year-round groundwater connection of the peat or bog body.
– Soil tillage and liming of soil with high humus contents that was originally near groundwater should be reduced to a minimum.

– The optimum use of such land is permanent grassland. The regulation of groundwater levels across all areas requires planning for the entire area.

Performance measurement

The reactivation of wetlands is particularly effective in the presence of mineralisation-intensive low moorland peat that is countering a strongly increased release of nitrate due to drainage. With sufficient groundwater retention, nitrate leaching can even be reduced to zero. This requires that the groundwater can be raised all year round to the base of the peat body. Such a measure is usually only possible in purely grassland areas.

▶▶ Biodiversity

F. Kaemena and L. Bastiaansen-Aantjes

Grasslands are areas covered by grass-dominated vegetation with little or no tree cover. Various types of grasslands exist in Europe, from desert-like in south-eastern Spain to steppes and dry grasslands, humid and damper grasslands and meadows that are often on deeper and more fertile soils, and finally lowland and mountain areas that dominate in the north and north-west (EC, 2008).

Most European grasslands can be defined as 'semi-natural' because they have developed through natural processes over long periods of grazing by domestic stock, cutting and even deliberate light burning regimes; others may have originated from sown and grass leys aimed at producing forage for livestock. In almost all cases, they are modified and maintained by human activities, mainly through grazing and/or cutting regimes (Turbé *et al.*, 2010). In this section, the term 'grasslands' includes meadows, steppes and grasslands managed (grazing, cutting, burning) with variable intensity. There is a large overlap with agro-ecosystems, which are covered in the corresponding section.

Table 5.3. Surface area (km^2) of grassland ecosystem in 1990, 2000 ad 2006.

	1990	2000	2006
Pastures	292 264	290 903	289 711
Natural grasslands	77 308	75 795	75 514
Total	369 572	366 697	365 224

Note: geographical coverage: EU except Greece, Finland, Sweden and the United Kingdom.
Source: Corine land cover.

Introduction

Biodiversity includes all living organisms found on land and in water. All those living organisms have a role in the 'fabric of life'. All the species, from the smallest bacteria in the soil, to the largest whale in the ocean. Biodiversity consists of four basic building blocks: genes, species, habitats and ecosystems (see box below).

Genes. Genes are the basic building blocks of life. They determine the characteristics of all living organisms. Maintaining genetic diversity by conserving species and varieties is a cornerstone of nature conservation.

Species. Nearly two million species have been identified worldwide and it is estimated that these may represent only 20% of the total currently existing on Earth. Soils alone host over one quarter of all species. Apart from microorganisms, insects are the biggest and most varied group. Other large groups include fungi, plants, lichens and mosses. Compared to other continents, Europe and the EU have relatively few species, although many are only present in the region (i.e. they are endemic).

Habitats. Different species of plants and animals come together to form ecological communities in a given area or natural environment called habitats. A habitat includes physical factors such as soil, moisture, temperature and light. Habitats are formed in response to local environmental conditions such as soil type and climate. In Europe, human activities have played a major part in shaping and creating habitats that are of high biodiversity value (e.g. meadows).

Ecosystems. An ecosystem can include one or many different habitats. Healthy ecosystems help to maintain species and habitats as well as providing critical goods and services to human beings.

Source: EU 2010 Biodiversity Baseline. https://www.eea.europa.eu//publications/eu-2010-biodiversity-baseline-revision

The distribution of wildlife and the variety of landscapes in Europe are the product of complex interactions. The basic physical qualities of the rock, soil and climate provide underlying structure and continuing influence. But the majority of the detail has been shaped through millennia of natural processes and human activity, the history of land use and management and its associated impacts. Human activities are themselves driven by economic, social and environmental forces.

As a result of these interactions, which are particular to Europe, 'multifunctional landscapes' have developed in which traditional cultural practices sustain a range of economic, social and environmental services. Among many of those multifunctional landscapes are pastures and natural grasslands. Due to human activities, the biodiversity on those agricultural landscapes is under pressure.

Status and trends

The serious and continuing loss of Europe's biodiversity reflects the continuing decline in the ability of ecosystems to sustain their natural production capacity and perform regulating functions. For instance, healthy soil biodiversity is fundamental to maintaining and ensuring soil fertility and therefore production potential. The Secretariat of the Convention on Biological Diversity (CBD) has observed:

'The loss of biodiversity often reduces the productivity of ecosystems, thereby shrinking nature's basket of goods and services, from which we constantly draw. It destabilises ecosystems, and weakens their ability to deal with natural disasters such as floods, droughts, and hurricanes, and with human-caused stresses, such as pollution and climate change. Already, we are spending huge sums in response to flood and storm damage exacerbated by deforestation; such damage is expected to increase due to global warming' (CBD, 2010).

Grasslands are among the most species-rich vegetation types (up to 80 plant species/ sqm) in Europe and have considerable conservation value (Eriksson *et al.*, 2002; Poschlod and Wallis de Vries, 2002; Wallis de Vries *et al.*, 2002, in Vandewalle *et al.*, 2010). There are different types of meadow habitats: natural, semi-natural, calcareous, dry, mesophile and humid; this reflects the high diversity of grasslands. Most of these have been created, modified or maintained by agricultural activities.

Large areas of grassland have been lost in recent decades, causing severe fragmentation of the remaining habitat areas and a consequent drop in populations of certain species by as much as 20% to 50% across Europe (EC, 2008). Grasslands are key habitats for many species: plants, butterflies, reptiles and many birds as well as grazing mammals such as deer and rodents. The grassland production plant biodiversity is already mentioned in previous chapters. Therefore, this chapter focuses on animal biodiversity, in particular meadow birds and butterflies.

Meadow birds

Meadow birds have their origins in savannah steppes from outside Europe. However, they like the open grassland areas and have made grasslands their home. The general trend for all countries is that the number of meadow birds is decreasing, as shown in Figure 5.6.

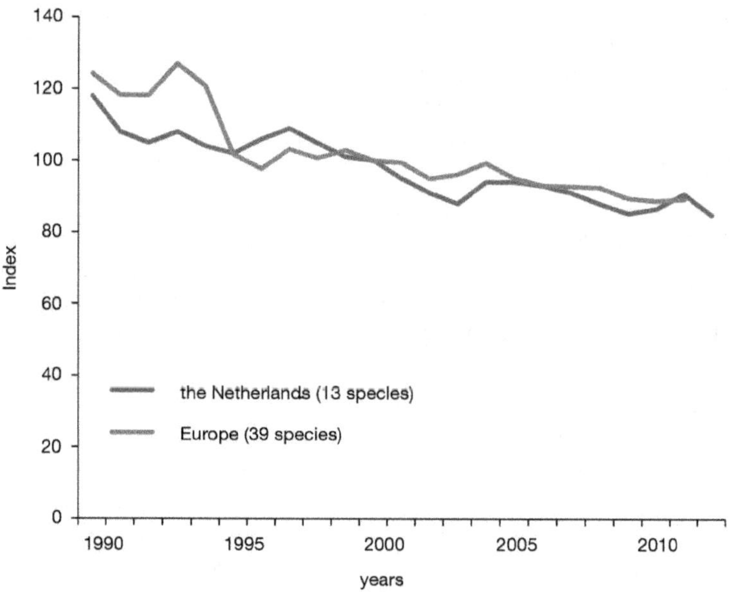

Figure 5.6. Farmland bord index for Europe (39 species) and The Netherlands (13 species). The index is set to 100 in year 2000.

Source: The Breeding Bird Monitoring Program (organised by SOVON in close collaboration with Statistics Netherlands and provinces and funded by the Dutch Ministry of EZ), European Bird Census Council (EBCC). –Entwicklung der Bestandsindizes für Vögel der Agrarlandschaft in Europa (39 Arten) und der Niederlande (13 Arten). Im Jahr 2000 steht der Index bei 100. http://www.birdnumbers2016.de/downloads/birdnumbers2016_example_manuscript.pdf.

Most populations are in heavy decline due to intensification of agricultural management and higher predation pressure. Intensification of agricultural management includes:
– Lowering of water tables for good horticultural conditions
– Larger machinery
– More mowing on the same dates

This is related to changes in land use and landscape (Source: http://www.altwym.nl/en.php/project/ ecological-monitoring/meadow-bird-ecology/)

Large breeding populations of black-tailed godwit, lapwing, redshank and oystercatcher can still be found with several other species. The status of birds by country is available on the website of the European Bird Census Council (http://www.ebcc.info/status-of-common-bird-monitoring-and-atlas-workin-single-states/). Additionally, the Euro Bird Portal offers a track and trace option of many bird populations throughout Europe (https://www.eurobirdportal.org/ebp/en/).

Of the 152 grassland bird species, 89 (59%) have an unfavourable conservation status in Europe (Birdlife International in Veen *et al.*, 2009). This is a slight deterioration compared to a decade ago, when the figure was at 81. Many of the now threatened species were formerly common in Europe, such as the lapwing (*Vanellus vanellus*), European starling (*Sturnus vulgaris*) and corn bunting (*Miliaria calandra*) (Tucker and Heath, 1994, in Veen *et al.*, 2009). It is important for young farmers to find out which species are threatened in their country and how they can help them.

Butterflies

Butterflies are special animals. During their life cycle, they transform from eggs to caterpillars (larvae), pupae and finally butterflies (Figure 5.7). They need particular conditions where they live; diversity of vegetation is key. Flowers are also important for butterflies to survive because they drink the nectar of the flowers. Meadow butterfly caterpillars mainly eat grasses and clovers.

To prevent predation, butterflies need bushes and vegetation between one and 1.5 meters high as a shelter. Butterflies also like high temperatures, so they seek out sunny, open spaces where there is no wind. It is important to create these spaces in grasslands.

European grassland butterfly abundance has declined by 60% since 1990, and this drop shows no sign yet of levelling off (EEA, 2009). Intensification in use and production across the relatively flat areas of Europe is the most important threat to butterflies.

For the yearly butterfly count, several documents are available per country. Figure 5.8 shows the counts for the United Kingdom.

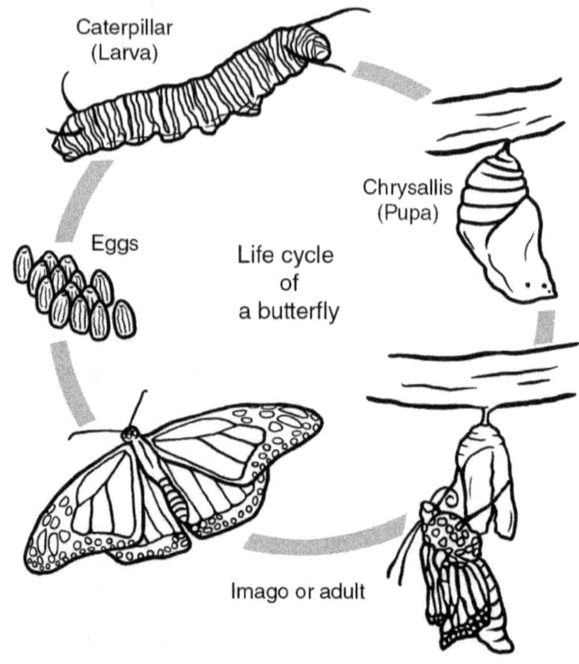

Figure 5.7. Butterfly life cycle.

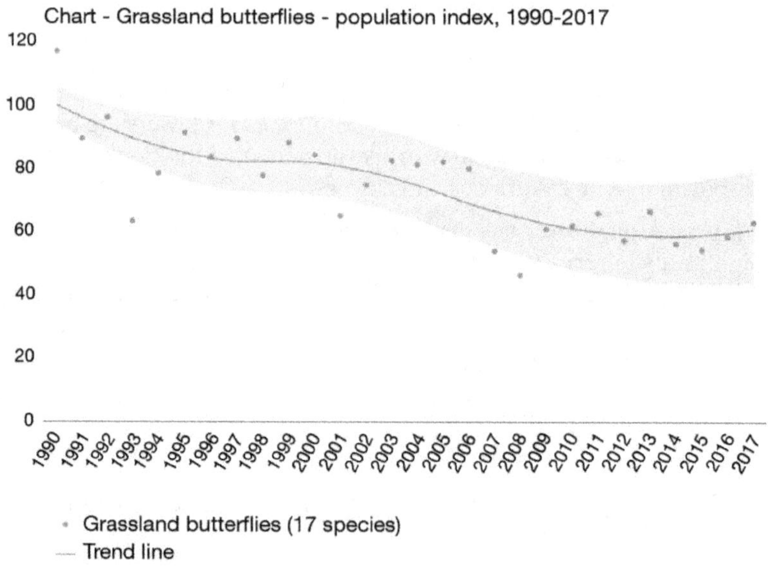

Figure 5.8. Grassland butterflies population index. 1990 = 100.

Biodiversity in Germany: a case study

Of the endangered species of ferns and flowering plants in Germany, about 40% grow mainly in grasslands. The dramatic decline in the number of species of wild bees and butterflies is directly related to the reduction in species-rich grasslands. The overwhelming significance of species-rich grasslands in water, soil and climate protection is often overlooked. A high diversity of grasses and herbs has a proven positive effect on groundwater quality and the carbon storage capacity of soil. In fact, grassland areas are the second most important carbon sink after marshlands, making the preservation of species-rich historical meadows and pastures hugely important for climate protection.

In recent years, there has been a dramatic reduction in species-rich grassland areas. A distinction can be drawn between the loss of permanent grasslands and the decline in the proportion of species-rich grassland areas. The most important causes of the reduction in the proportion of permanent grasslands from agricultural land areas are the intensification of agriculture (especially milk and meat production), as well as the increased cultivation of energy crops (especially corn) for biomass production. In low mountain regions, the abandonment of grasslands also plays an important role. The deterioration in the quality of species-rich grasslands (loss of species) observed in many places is also related to agricultural changes. The palpable changes in the growing season as a result of climate change has also left their mark on grasslands, as these changes impact the less-competitive species first and foremost.

Grasslands are indispensable components in multi-functional agriculture, and not only for agricultural production. They also have tremendous value for biodiversity, as recreational areas for the local population, and for a multitude of nature conservation and environmental concerns.

No other part of the world has such a wide variety of cultivated grassland ecosystems (Dierschke and Briemle 2002). Certain long-standing, extensively-used types of grasslands, such as the limestone grasslands, are among the most species-rich biotopes in Central Europe. A third of all indigenous ferns and flowering plants grow mainly in the grasslands (1,250 of 2,997 species assessed as belonging to a vegetation unit and as endangered). Of the endangered species of ferns and flowering plants in Germany, around 40% (or 822 species) are found in grasslands. Grasslands had their greatest diversity of species and communities in times of semi-extensive to semi-intensive land use, i.e. mainly from the 18th to the middle of the 20th century (Dierschke and Briemle 2002).

With their many structures and seasonal flowering sequences, grasslands provide a habitat for a great variety of animals, ranging from vertebrates such as birds and amphibians to the microorganisms of flowers and inflorescences, with very close interrelations between flora and fauna (Dierschke and Briemle 2002). Due to the enormous species spectrum and the large number of different sites, the conservation of grassland plays an essential role in achieving national, European and international biodiversity goals.

Lapwings breed in open, flat landscapes and prefer short grass.

The bird populations in cultivated areas in Germany are coming under increasing pressure – in agricultural areas, the number of birds has halved over the last 30 years. According to the results of the 2013 national bird protection report, the number of species with declines in the 12-year period has increased significantly compared to the 25-year period: over the short-term period, a third of all breeding bird species (84 species) show significant declines. The largest percentage of declining populations is accounted for by species in the open land and settlement areas.

Most bird species that breed on fields, meadows and pastures are clearly declining in numbers due to the high intensity of agricultural activity. In species such as the lapwing and the black-tailed godwit, which breed predominantly in wet meadows on the ground, population losses have persisted for decades. The population of the lapwing has shrunk to a quarter of what it was over the last 20 years, while that of the black-tailed godwit has halved. During the same period, both species also show declining populations throughout Europe.

The dusky large blue is dependent on grassland areas with burnet.

Agricultural areas in Europe have now lost about half of their birds. In total, 300 million fewer birds than just 30 years ago live in the open cultivated area of the European Union today. The European Farmland Bird Indicator has also fallen by more than 50 per cent since 1980. The most important causes for the threat to important bird species are the loss of breeding and food habitats due to increasingly intensive agriculture and the drainage of agricultural land. The tillage and drainage of grasslands have significantly reduced the area of habitat for breeding birds in wet meadows. The conversion of meadows and pastures into arable land for the cultivation of bioenergy crops has drastically aggravated the situation in recent years.

With the decline in grassland, insects such as bees and butterflies, which depend on an abundant supply of flowers and nectar, lose their food source and habitat. The current red list of invertebrates (Binot-Hafke *et al.*, 2011) shows that the downward population trend has continued, in particular, for butterfly species found on alkaline grasslands and dry grasslands and bees found in hay meadows, alkaline grasslands and heaths. In addition, there is the loss of ecotone in agricultural areas, which directly harms small mammal and amphibian (e.g. common frogs) populations in wet grasslands and the populations of species in or in contact with fresh and dry grasslands, such as the endangered common hamster and the sand lizard. In the current red lists (Binot-Hafke *et al.*, 2011), species linked to habitats rich in small structures, such as orchards, also show downward trends (e.g. bees, ants).

Grassland areas with rich biodiversity are the habitat for numerous endangered animal and plant species.

Chapter 6

Quality of grassland-based animal products

The text below is based on Couvreur S. *et al.*, *Les Prairies au service de l'Elevage - Comprendre, gérer et valoriser les prairies*, Educagri, 2018.

▸▸ Milk and dairy product quality

Dairy product quality is highly dependent on the initial composition of the milk and on the same factors of variation on the farm as the milk itself. Milk quality depends on many factors: species, breed, stage of lactation, animal health condition and feeding. When feeding is based on grazed grasslands or grass-based forage, the useful components of the milk change: proteins and fat as well as fatty acids, vitamins, pigments, phenolic compounds and terpenes. The effect of grass-based feeding on milk quality particularly depends on how it is provided (grazed grass or conserved forage), its overall proportion in the whole ration, its vegetative stage, its botanical composition and its supplementation.

▸▸ Effect of grazed grass on milk composition

Compared to maize, grazed grass is very rich in soluble sugars, α-linoleic acid (C18:3n-3)-rich lipids (linoleic acid or C18:2n-6 for maize) and in minor components (pigments, vitamins and terpenes) whose quantity varies depending on the botanical richness. This difference in composition has a big impact on milk composition.

For example, an exclusively spring grass-based ration, compared to maize, will have:
– A similar or slightly higher protein level, and frequently, a lower fat level. However, the fat level increases when the diet is exclusively pasture-based compared to pasture and concentrates. With regard to fatty acid variation:
  • Saturated fatty acid levels decrease. The effects are higher in cow milk than for sheep and goat milk.
  • C18:2n-6 level decreases while C18:1cis9 and C18:3n-3 levels increase with the same variation between species as mentioned above.
  • Total trans fatty acid levels increase in cow and sheep milk, while they do not vary much in goat milk. This can be explained by differences in metabolism as well as because rations contain high grass-based conserved forage levels and lipid supplementation.
– A variation in pigments and vitamin levels:
  • B-carotene levels (pigment, precursor to vitamin A synthesis) increase (only in cow milk)
  • Vitamin A and E levels increase in cow, sheep and goat milk
  • Vitamin B12 level decreases
  • Vitamin B2 and B9 levels increase
– An increase in phenolic compounds and terpene levels.

The degree of variation in these levels will depend on the grass species and the vegetation stage of the grassland. The more diverse the botanic composition (legumes, herbs), the more important the effects on the phenolic compounds and terpene levels. The later the vegetative stage, the more mitigated the effects on fat and protein, fatty acids, pigments and vitamin A levels, depending on the different leaf and stem quantities. Finally, grazing management also impacts milk composition. When grazing managements favour quantity and consistent quality grass consumption, greater variations appear in fatty acid levels. When grazing management is rationed out with high stocking rates on diversified grassland, terpene levels can be higher. For cows, an increase in the share of grass in a maize silage-based ration will result in a decrease in fat, an increase in protein up to certain a threshold and a modification of the fatty acid composition.

▶▶ Effect of the grass conservation type on milk composition

Forage composition depends on the methods used to harvest and conserve the grass, which in turns impacts milk composition. The effects of forage on milk fat and protein composition, fatty acid profiles and pigments and carotene content are as follows, in order of highest to lowest: grazed grass > fresh grass to the trough > hay > grass silage > grass haylage. The order can vary depending on the harvesting conditions of the forage. Forage conserved with a highly diversified composition in terms of the botanical aspects still remains rich enough to have an effect on the phenolic compounds and terpene levels in milk.

Dairy product quality

Dairy products are made from milk. Their quality is determined according to the following properties:
– Organoleptic (texture, taste, flavours, colour etc.)
– Nutritional
– Technological (yield)
– Functional (spreadability, melting, flowing etc.)

Grass-based diets, which modify the fat, vitamin, pigment, phenolic compound and terpene composition of milk, have comparable effects on butter and cheeses.
– The texture of butter (spreadability and melting) and cheese at a given temperature is explained by the ratio of solid to liquid fat. This ratio varies according to the fatty acid composition, and depends in particular on the two main components: C16:0 (which has a firming effect) and C18:1cis9 (which has a softening effect). Thus, by increasing C18:1cis9 content and decreasing C16:0 content, a grass-based diet increases the spreadability and melting in the mouth compared to maize silage-based diets. As soon as there is more than 60% of spring grass brought to the trough, this effect becomes significant. The more grass has been produced at a leafy stage, the greater the effect on texture.
– The colour of butter and cheese is directly related to the β-carotene content. Compared to maize silage, the grass is richer in carotenoids and therefore produces yellower products (only in cow milk products). In the production of preserved grass forage, the drying time on the ground reduces the pigment content. The effects of forage on colour are as follows, from highest to lowest: grazed spring grass > grazed grass in summer (bushy) > ventilated hay > grass silage > haylage > hay dried on the ground.
– The flavour of butter and cheese is the result of many interactions between the initial composition of milk and its evolution during processing and ripening. Compared to hay, grass gives products a stronger flavour. The presence of natural antioxidants from grasslands (vitamin A and E) decreases the oxidation defect on taste (a metallic or fishy taste). As for other quality characteristics, these effects are greater when grass is brought fresh, leafy and in large quantity in the ration.

Meat quality

Diet management and the type of feed have varying effects on the organoleptic and nutritional quality of the meat. The type feed has little effect on the tenderness of beef (pasture, forage kept dry or wet). In sheep production, fattening systems (on grass or in buildings) have no significant effect on the tenderness of the meat either. Livestock farming practices with regard to grazing management in cattle and sheep do not have an impact on the juiciness of the meat.

Finishing sheep and cattle on grass does change the flavour of the meat. This is explained by differences in fatty acid profiles caused by the presence of grass in the ration (a phenomenon comparable to what has been described for milk composition). During cooking, fatty acids produce different volatile compounds that give meat a 'pastoral' flavour.

Pasture feeding has an effect on the colour and the fat of the meat. Cattle and sheep reared on grass produce a darker and more red meat than animals fed in the trough (see photos in Figure 6.1), due to a higher content of a pigment called myoglobin. Additionally, the fat of sheep and cattle meat is more yellow (especially when grasslands are grazed at the leafy stage), which is explained by a transfer of a larger amount of β-carotene ingested to the adipose tissue.

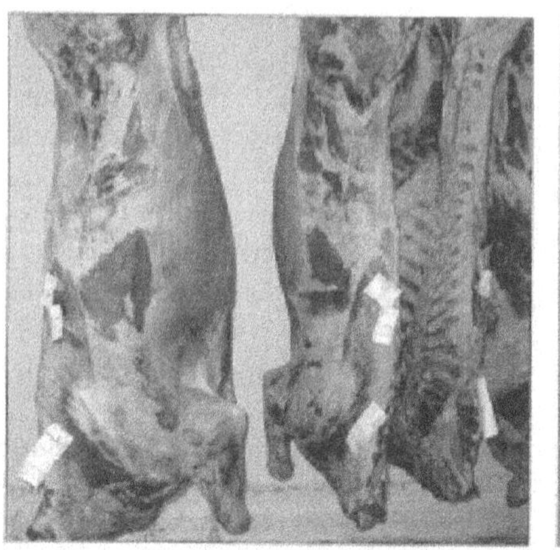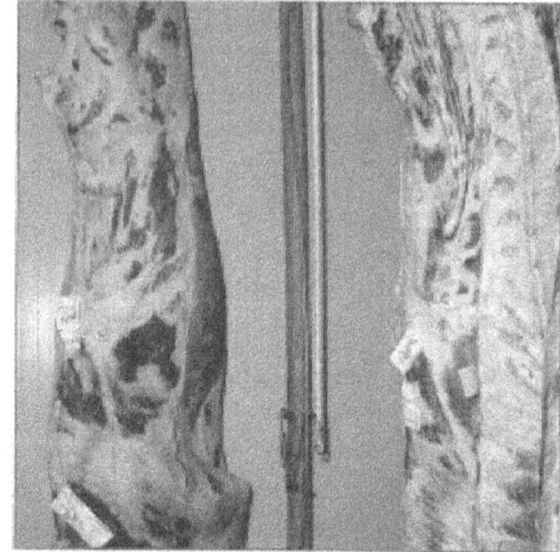

Figure 6.1. Comparison of carcasses from animals fed with grass (on the left) or with maize silage (on the right).

The nutritional quality of meat produced by animals that are grassland grazed or given preserved forage is higher, and especially the fatty acid composition due to increased C183n-3, C18: 1trans11, C18: 2cis9trans levels and decreased saturated fatty acid content. The nutritional quality is preserved thanks to the transfer of vitamin E from the grass by protecting the good fatty acids from being oxidised.

Case study: Farmer testimonial on the effect of grassland use on cheese – GAEC des Violettes (Puy-de-Dôme, France)

A total of 47 dairy cows of Montbéliarde and Abondance breeds are milked on this farm, which has 57 ha of permanent grasslands located at 1,000 m altitude. The milk is processed into Saint-Nectaire cheese on the farm. The calving period extends throughout the year to ensure regular cheese production and sale. The dairy cow ration grass-based, with grazing from spring to autumn and preserved forage (hay/ leftovers) in winter. Concentrates are distributed during the year but never beyond 230 g/l of milk. The farmer uses rotational grazing and changes plots every three days. A hay supply is maintained throughout the year, but consumption is very low during the grazing period.

Farmer viewpoint	Cheesemaker viewpoint
Longer transitions to manage the cheeseability of the milk	
"As we make cheese, we pay close attention to transitions. We used to make abrupt transitions and we needed a perfect mastery and adaptation of the cheese technology. To avoid this, we now make long transitions. When we turn the cows out for grazing or back into the barn, the transition is about one month. When we change a feed, the transition is also long to limit sudden changes in the milk quality."	"When we turn the cows out for grazing, the grass is rich and the vegetation is optimal. Rates increase at this time. In terms of cheese processing, care must be taken to squeeze the cheese to prevent it from crashing. We need to be very reactive because problems can appear two to three weeks later with an impact on several batches. When we turn the cows back into the barn, we have the opposite situation. Because the milk is not as rich, we must be careful not to over-tighten the cheeses."
Hay supply during the grazing period to better manage cheesability	
"In addition to paying attention to the transitions, I always leave a little hay in the summer to stabilise the belly."	"The year-round supply of hay – and therefore a little fibre – leads to fewer fits and starts on cheeses. During the grazing period, the plot changes to the amount of milk to be processed. At the end of grazing of a plot, the quantity of milk decreases: the vat of milk is not as full, which then makes processing difficult. When the tanks are filled to less than three-quarters full, processing is more difficult."
Overall effect of the use of grass in the ration on cheeses	
"As we make cheese, we mainly look at the rates (fat and protein) more than the quantity of milk. We work with what forage and grass we have: we have to work with the available grass, especially with permanent grasslands. We see many differences in the milk (in quantity) according to the grasslands but the cheesemaker adapts. We warn each other when there are changes."	"Feed has an effect on cheeses: in winter the curds are whiter and in summer they yellower. The flavours will change too. The grass is what gives the product its typical flavour. The cheese-making process is more complex with grass, but it is part of our image and the cheeses are tastier. And if we plan ahead and adapt, everything goes well. We work with the living world: it changes constantly."

References

Anonymous, 1997. DLG-Futterwerttabellen - Wiederkäuer. 7., erweiterte und überarbeitete Auflage, DLG-Verlag, Frankfurt am Main.

Archambeaud M., Thomas F., 2016. *Les sols agricoles comprendre, observer, diagnostiquer*, Editions France Agricole, Paris.

Bachelier G., 1978. La faune des sols son écologie et son action, ORSTOM, Paris.

Barenbrug Agriseeds, 2019. Maximising ryegrass growth: www.agriseeds.co.nz/sheep-beef-deer/pasture-management/maximising-ryegrass-growth 2019-02-16.

Baumgartner, 2006. Richtlinie für die Sachgerechte Düngung. Anleitung zur Interpretation von Bodenuntersuchungsergebnissen in der Landwirtschaft, 6. Auflage.

Calvet R., Chenu C., Houot S., 2015. *Les matières organiques des sols*. Editions France Agricole, Paris.

Couvreur S., 2018. *Les prairies au service de l'élevage*. Educagri éditions, Dijon

Daccord R., Wyss U., Jeangros B., Meisser M., 2007. Bewertung von Wiesenfutter. Nährstoffgehalt für die Milch- und Fleischproduktion. Zürich: AGFF (AGFF Merkblatt, 3).

DairyNZ, Grass Silage: https://www.dairynz.co.nz/feed/supplements/grass-silage/

De Brogniez D., Ballabio C., Stevens A., Jones R.J.A., Montanarella L., van Wesemael B., 2015. A map of the topsoil organic carbon content of Europe generated by a generalized additive model. *European Journal of Soil Science*, 66, 121–134.

DEFRA, 2018. Department for Environment Food and Rural Affairs. https://www.gov.uk/guidance/grassland-derogations-for-livestock-manure-in-nitrate-vulnerable-zones

Delisle C., 2010. Dossier Luzerne. La Luzerne reine des fourrages. *Réussir bovins viande*, 167, 14-33.

Deprez B., Parmentier R., Lambert R., Peeters A., 2005. Prairies temporaires pour des systèmes agricoles durables adaptés aux fermes mixtes de la moyenne Belgique. Rapport final, Laboratoire d'Ecologie des Prairies, 133 p.

Dietl W., Lehmann J., Jorquera M., 1998. Wiesengräser. Landwirtschaftliche Lehrmittelzentrale, Zollikofen.

Dietl W., Jorquera M., 2003. Wiesen- und Alpenpflanzen. Erkennen an den Blättern, Freuen an den Blüte. Agrarverlag, Leopoldsdorf.

Dietl W., Lehmann J., Jorquera M., Scotton M., 2012. Le graminacee prative. Patron Editore, Bologna.

Dinnes D.L., Karlen D.L., Jaynes D.B., Kaspar T.C., Hatfield J.L., Colvin T.S., Cambardella C.A., 2002. Nitrogen management strategies to reduce nitrate leaching in tile-drained Midwestern soils. *Agronomy Journal*, 94, 153-171.

DLG, 1997. DLG-Futterwerttabellen – Wiederkäuer. 7., erweiterte und überarbeitete Auflage. Frankfurt am Main: DLG-Verlag.

DRAAF-Bretagne, 2015. Nitrogen fertilization of grasslands: methodology. http://draaf.bretagne.agriculture.gouv.fr/IMG/pdf/GREN_annexe8-1_prairies_09_03_2017_cle874215.pdf

Dunière L., Sindoub J., Chaucheyras-Durand F., Chevallierd I., Thévenot-Sergente D., 2013. Silage processing and strategies to prevent persistence of undesirable microorganisms. *Animal Feed Science and Technology*, 182, 1–15.

Elsäßer M., 2009. Gülledüngung im Grünland. Merkblätter für die Umweltgerechte Landbewirtschaftung, Nr. 26.

EU, 2010. Biodiversity Baseline. Available at: https://www.eea.europa.eu//publications/eu-2010-biodiversity-baseline-revision

FAOStat, 2016. Global inputs of fertilizers. Available at: http://www.fao.org/faostat/en/#data/RFN, accessed 19.12.2018.

Fertilizing Guideline of Northern Germany, 2018. Richtwerte für die Düngung. Landwirtschaftskammer Schleswig-Holstein.

Fourrages Mieux, 2008. Legumes: Associations, simple mixing, complex mixing: How to make the right choice?

Frame J., Laidlaw A.S., 2014. Improved grassland management. New edition. Crowood, New York.

Genever L., McConnell D., 2014. Growing and Feeding Lucerne. AHDB Beef & Lamb, AHDB Dairy, United Kingdom, Technical leaflet, 20 pp.

Godfray H.C., Beddington J.R., Crute I.R., Haddad L., Lawrence D., Muir J.F., Pretty J., Robinson S., Thomas S.M., Toulmin C., 2010. Food Security: The challenge of feeding 9 billion people. *Science,* 327, 812-818.

Good A.G., Beatty P.H., 2011. Fertilizing nature: A tragedy of excess in the commons. *PLoS Biology,* 9, 1-9.

Haynes R.J., Williams P.H., 1993. Nutrient Cycling and soil fertility in the grazed pasture ecosystem. *Advances in Agronomy,* 49, 119-199.

Hou Y., Velthof G.L., Oenema O., 2015. Mitigation of ammonia, nitrous oxide and methane emissions from manure management chains: A meta-analysis and integrated assessment. *Global Change Biology,* 21, 1293-1312.

Hubbard C.E., 1992. *Grasses: A Guide to Their Structure, Identification, Uses and Distribution.* Penguin Press Science, London.

Hubbard C.E., Boeker P., 1985. *Gräser: Beschreibung, Verbreitung, Verwendung.* 2., überarbeitete und ergänzte Auflage. Stuttgart: Ulmer Verlag (Uni-Taschenbücher, 233).

Huyghe C., De Vliegher A., van Gils B., Peeters A., 2014. *Grasslands and Herbivore Production in Europe and Effects of Common Policies.* Editions Quae, Versailles, France, 320 pp.

Jordbruksverket, 2018. Fertilisation. Available at: https://www.jordbruksverket.se/swedishboardofagriculture/engelskasidor/crops/plantnutrients.4.6621c2fb1231eb917e680003205.html

Klapp E., Opitz von Boberfeld W., 1990. *Taschenbuch der Gräser.* 12. Auflage. Berlin, Verlag Paul Parey, Hamburg.

Klapp E., Opitz von Boberfeld W., 2004. *Kräuterbestimmungsschlüssel für die häufigsten Grünland- und Rasenkräuter.* Blackwell Wissenschafts-Verlag, Berlin.

Klapp E., Opitz von Boberfeld W., 2011. *Gräserbestimmungsschlüssel für die häufigsten Grünland- und Rasengräser.* Blackwell Wissenschafts-Verlag, Berlin.

Klocker H., Prünster T., Peratoner G., Gauly M., 2017. BRING, Beratungsring Berglandwirtschaft. Leitfaden Düngung Grünland, 01/2017.

Klotz C., Cassar A., Florian C., Figl U., Bodner A., Peratoner G., 2012. Selenium fertilization of grassland: effect of frequency and methods of application. Grassland – a European Resource. *Grassland Science in Europe,* 17, 367-370.

Knoden D., Lambert R., Nihoul P., Stilmant D., Pochet P., Crémer S., Luxen P., Reasonable Fertilization of Grasslands, 2007. Agriculture booklet n°15, Wallonia, Ministry of Walloon Region. Available at: http://www.fourragesmieux.be/Documents_telechargeables/Livret_fertilisation.pdf

Kung L., Shaver R., 2001. Interpretation and Use of Silage Fermentation Analysis. *Focus on Forage,* 3, 13,1-5.

Landolt E., Bäumler B., Erhardt A., Hegg O., Klötzli F., Lämmler W., Nobis M., Rudmann-Maurer K., Schweingruber F.H., Theurillat J.-P., Urmi E., Vust M., Wohlgemuth T., 2010. Flora indicativa: ökologische Zeigerwerte und biologische Kennzeichen zur Flora der Schweiz und der Alpen. Bern: Haupt.

Lasar A., 2017. Climate efficiency of milk production; Systemanalyse Milch – Hintergründe für die Praxis, 59 pp.

Marley C.L., Fychan R., Fraser M.D., Sanderson R., Jones R., 2007. Effects of feeding different ensiled forages on the productivity and nutrient-use efficiency of finishing lambs. *Grass and Forage Science*, 62, 1-12.

MEA, 2005. Ecosystems and Human Well-being: Current State and Trends, Volume 1. 901 pp.

Ministry of Agriculture, Food and Rural Affaires, Ontario, 2006. Pasture Grasses identifier. Available at: http://www.omafra.gov.on.ca/english/livestock/beef/facts/06-095.htm 2019-02-16.

NC State Extension Publications: https://content.ces.ncsu.edu/forage-conservation-techniques-silage-and-haylage-production

ND, Nitrates Directive, 1991. (91/676/EWG).

Nevens F., Reheul D., 2003. Effects of cutting or grazing grass swards on herbage yield, nitrogen uptake and residual soil nitrate at different levels of N fertilization. *Grass and Forage Science*, 58, 431-449.

Olesen J.E., Trnka M., Kersenbaum K.C., Skjelvag A.O., Seguin B., Peltonen-Sainio P., Rossi F., Kozyra J., Micale F., 2011. Impacts and adaptation of European crop production systems to climate change. *European Journal of Agronomy*, 34, 96-112.

Oregon State University, 2019. Forage Information System: https://forages.oregonstate.edu/regrowth/how-does-grass-grow/grass-structures 2019-02-16.

Peeters A., Wezel A., 2017. Agroecological Principles and Practices for Grass-based Farming Systems. Chapter 11. In: Wezel A. (Ed.) *Agroecological Practices for Sustainable Agriculture*. World Scientific, Connecting Great Minds, 293-354.

Peeters A., Beaufoy G., Canals R.M., De Vliegher A., Huyghe C., Isselstein J., Jones G., Kessler W., Kirilov A., Mosquera-Losada M., Nilsdottir-Linde N., Parente G., Peyraud J.L., Pickert J., Plantureux S., Porqueddu C., Rataj D., Stypinski P., Tonn B., van den Pol-van Dasselaar A., Vintu V. , Wilkins R.J., 2014. Grassland term definitions and classifications adapted to the diversity of European grassland-based systems. *Grassland Science in Europe*, 19, 743-750.

Peratoner G., Figl U., Florian C., Senoner J.L., Ros G. de, 2015. Studio dei costi di produzione del foraggio nella Provincia di Bolzano (BLW-gw-11-1) Relazione finale di progetto. Unter Mitarbeit von N. Zenleser, P. Steger, R. Großrubatscher und G. Tschurtschenthaler. Vadena/Pfatten: Land- und Forstwirtschaftliches Versuchszentrum Laimburg.

Quakernack R., Pacholski A., Techow A., Herrmann A., Taube F., Kage H., 2012. Ammonia volatilization and yield response of energy crops after fertilization with biogas residues in a coastal marsh of Northern Germany. *Agriculture, Ecosystems and Environment*, 160, 66-74.

Resch R., Guggenberger T., Gruber L., Ringdorfer F., Buchgraber K., Wiedner G., Kasal A., Wurm K., 2006. Futterwerttabellen für das Grundfutter im Alpenraum. In: *Der fortschrittliche Landwirt*, 84 (24), S. 1–20.

Schilling G., 2000. *Pflanzenernährung und Düngung*. Eugen Ulmer GmbH & Co. Stuttgart.

Schubert S., 2006. *Pflanzenernährung Grundwissen Bachelor*. Eugen Ulmer KG. Stuttgart.

Schulte R.P.O., Lanigan G., Gibson M., 2011. Irish agriculture, greenhouse gas emissions and climate change: opportunities, obstacles and proposed solutions. Teagasc, 92 pp.

Smith K.A., Beckwith C.P., Chalmer A.G., Jackson D.R., 2002. Nitrate leaching following autumn and winter application of animal manures to grassland. *Soil Use and Management*, 18, 428-434.

Smith P., Olesen J.E., 2010. Synergies between the mitigation of, and adaptation to, climate change in agriculture. *Journal of Agricultural Science*, 148, 543-552.

Soussana J.F., Lemaire G., 2014. Coupling carbon and nitrogen cycles for environmentally sustainable intensification of grasslands and crop-livestock systems. *Agriculture, Ecosystems and Environment,* 190, 9–17.

Steinshamn H., 2010. Effect of forage legumes on feed intake, milk production and milk quality - a review. *Animal science papers and reports,* 28, 3, 195-206.

TEAGASC, 2018. Grassland N fertiliser advice for dairy grazing. Available at: https://www.teagasc. ie/crops/soil--soil-fertility/grassland/

The Seed Biology Place, 2019. http://www.seedbiology.de/structure.asp#caryopsis 2019-02-16.

Trott H., Wachendorf M., Ingwersen B., Taube F., 2004. Performance and environmental effects of forage production on sandy soils. I. Impact of defoliation system and nitrogen input on performance and N balance of grassland. *Grass and Forage Science,* 59, 41-55.

Van den Pol-van Dasselaar A., Vellinga T.V., Johansen A., Kennedy E., 2008. To graze or not to graze, that's the question. *Grassland Science in Europe,* 13: 706-716.

Van den Pol-van Dasselaar A., Bannink A., 2014. Qualitative overview of mitigation and adaptation options in livestock systems. *Grassland Science in Europe,* 19, 119-121.

Van den Pol-van Dasselaar A., Aarts H.F.M., De Caesteker E., De Vliegher A., Elgersma A., Reheul D., Reijneveld J.A., Vaes R., Verloop J., 2015. Grassland and forages in high output dairy farming systems in Flanders and the Netherlands. *Grassland Science in Europe,* 20, 3-11.

Van den Pol-van Dasselaar A., Becker T., Botana Fernández A., Hennessy T., Peratoner G., 2018a. Social and economic impacts of grass based ruminant production. 27th General Meeting of the European Grassland Federation, Cork, Ireland.

Van den Pol-van Dasselaar A., Chabbi A., Cordovil C., De Vliegher A., Die Dean M., Hennessy D., Hutchings N., Klumpp K., Koncz P., Kramberger B., Newell Price P., Poilane A., Richmond R., Rocha Correa P., Schaak H., Schönhart M., Sebastiá M.T., Svoboda P., Teixeira R., van Eekeren N., van Rijn C., 2018b. EIP-AGRI Focus Group Grazing for Carbon, Final report, 32 pp.

Van Grinsven H.J.M, Spiertz J.H.J., Westhoek H.J., Bouwman A.F., Ersiman J.W., 2014. Nitrogen use and food production in European regions from a global perspective. *Journal of Agricultural Science,* 152, 9-19.

Vergé X.P.C., De Kimpe C., Desjardins R.L., 2007. Agricultural production, greenhouse gas emissions and mitigation potential. *Agricultural and Forest Meteorology,* 142, 255-269.

Voigtländer G., Jacob H., 1987. *Grünlandwirtschaft und Futterbau.* Verlag Eugen Ulmer, Stuttgart.

Wageningen UR Livestock Research, 2014. Manure. A valuable resource.

Wba, Wissenschaftlicher Beirat für Agrarpolitik, 2016. Klimaschutz in der Land- und Forstwirtschaft sowie den nachgelagerten Bereichen Ernährung und Holzverwendung. Gutachten, Juli 2016.

Westerlind E., Svanäng K., af Geijersstam L., 1997. Artkaraktärer. Plantor av vallväxter. Sveriges lantbruksuniversitet. Institutionen för växtodlingslära. Kompendium. 17 pp.

Characteristics
of individual countries

Sweden

Nilla Nilsdotter-Linde, Eva Spörndly and Rolf Spörndly

▸▸ Introduction

In Sweden, short-term leys incorporated into arable crop rotations are the main forage crop, unlike the perennial forage swards farther south in Europe. Short-term leys and green fodder crops are the most widely grown crop type in Sweden (43% of arable land in 2017). When semi-natural grasslands are included, 51% of agricultural land is covered by grass. Due to the increasing price of concentrate, having regionally produced high-quality fodder is of fundamental importance for most producers. Perennial crops have a number of positive environmental effects in terms of decreasing nitrate leaching and nitrous gases, increasing carbon sequestration, improving soil structure and crop rotations etc. Swedish silage-based production and utilisation has been described previously (Spörndly and Nilsdotter-Linde, 2011). Here we provide a rewritten and updated description, including new developments in recent years.

▸▸ Climate conditions for grass and legumes in Sweden

Growth and development are influenced by temperature and light (light intensity and day length). The combination of temperature, insolation and day length is unique in Scandinavia/Fennoscandia and neighbouring parts of Russia. Insolation is not a limiting factor for growth, since most crops are C_3-plants. Temperature conditions are favourable in Scandinavia despite its northerly position, with all the Nordic countries except southern Denmark being above 55°N. Three Nordic capitals, Helsinki, Stockholm and Oslo, are located at roughly the same latitude as Anchorage, Alaska. However, thanks to the Gulf Stream, which influences winter temperatures especially, the climate is temperate, with a combination of favourable summer temperatures and long days. Despite the relatively low sun height, total global insolation is high due to the long day length.

Given the short growing season (day and night mean temperature >5°C), which ranges from 150 days in the far north to 240 days in southern coastal areas, the grazing season in Sweden is short and conserved forage on an annual basis accounts for about 50% of the total dairy cow rations, while the corresponding figure for pasture is only approximately 10%. However, for beef cattle, growing heifers and sheep, silage is generally the main feed in winter and pasture is the main feed in summer, when grazing is possible for three to five months depending on geographical location. In a change during recent years, an increasing number of horses are now fed silage and many horses are kept on pasture in summer.

▸▸ Feeding ruminants and horses up to the 1960s

The grazing season in Sweden is May to October in the south and June to August in the north. Traditionally, grazing was concentrated to areas not suitable for ploughing and cropping. Due to the accumulation of water and nutrients in the soil after six to nine months of winter and the long daylight periods in the spring (17 hours in the south and 22 hours in the north on 1 June), the grass growth rate is much faster in the early growing season. Therefore, in the past suitable grassland areas were excluded from grazing and used for production of hay for the coming winter. Hay was cut as a single cut in the end of June in the south and one month later in the north. The area was then used for grazing later on in the summer, when the decreasing growth rate required extended areas for grazing.

Haymaking was a time-consuming process that included cutting, raking, filling hay racks and finally transport to the hay barn. This produced feed that was adequate for horses in moderate work or very low-producing dairy cows. However, supplementation with concentrates was always needed. Barley and oats were usually grown for this purpose. Faba beans were sometimes used as a protein supplement but, in the 1950s, imported protein sources such as coconut cake, groundnut cake, cottonseed cake and soybean meal became very common.

From the 1950s on, the Swedish state invested significant resources in rationalising agricultural production. The main goal was to decrease the need for labour on farms in order to supply the fast-growing industries with more labour. In the 1960s, this resulted in farm mechanisation and amalgamation, and the number of farms decreased from over 200,000 dairy farms in 1960 to fewer than 85,000 in 1970. The decline in the number of farms is still continuing, and dairy farms are now down to 3,600, as shown in Figure SE1. However, total milk production in Sweden decreased by only 10% in the past 25 years, due to larger farms and higher yield per cow.

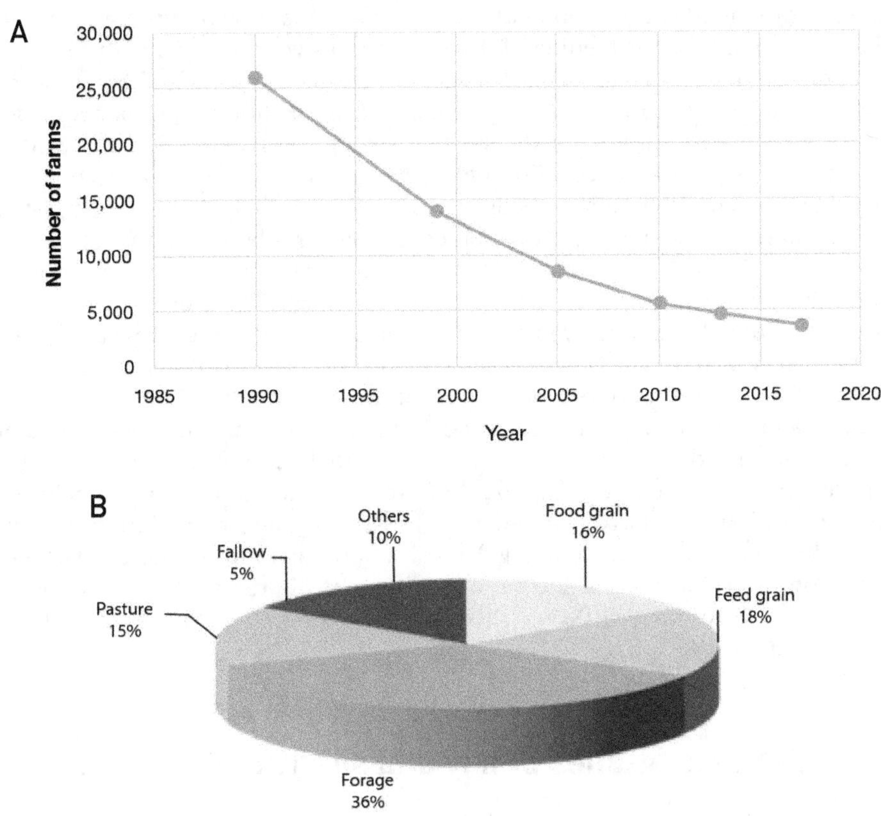

Figure SE1. Number of dairy farms in Sweden 1990–2017 (A) and use of agricultural land in Sweden 2017 (B) (Swedish Board of Agriculture, 2018a).

▶▶ Use of pasture: historical development

Historically, pasture was an important feed asset and cattle, sheep and horses were kept in wooded areas or on non-cultivated pastures in summer. As productivity increased with mechanisation and the introduction of mineral fertilisers, the most productive animals, such as dairy cows, were gradually moved to temporary grasslands near the milking parlour and growing stock was kept on semi-natural pastures. When interest in biodiversity started to increase in the 1970s, the importance of semi-natural pastures for biodiversity was recognised and the first subsidies to maintain the biodiversity values of these pastures were launched. These subsidies were gradually expanded and are now an important part of the production system for rearing heifers, steers, suckler cows with offspring and sheep.

In conjunction with the increase in farm size and the rationalisation of milk production that took place the 1960s and 1970s, grazing of dairy cows in particular started to decrease. This was especially apparent in northern parts of Sweden, where advisory organisations began recommending that farmers abandon outdoor grazing and turn to year-round indoor feeding of dairy cows. This was mainly due to the short grazing season in the region, combined with increasing farm size and higher levels of milk production. Efforts to organise grazing for large herds were not regarded as worthwhile or economical for the short grazing season, which sometimes lasts only two months. This practice started to spread southwards and gain acceptance. However, in the late 1970s a discussion started among farmers, the advisory system and the general public about production conditions in relation to animal welfare considerations. This attracted considerable interest among the public and resulted in new legislation to promote animal welfare on farms. Thus, in 1980 a law was passed stipulating that animals must be given the opportunity to express their natural behaviour, which for cattle, sheep and goats meant that they had to be given the opportunity to graze outdoors in summer. With the exception of bulls and calves younger than six months, the law required non-lactating cattle and sheep to be on pasture 24 hours a day. For dairy cows, the law initially required that they should be put to pasture between two milkings. However, since the law was launched in 1980, the requirements have been adapted to new production conditions such as the introduction of automatic milking. However, although the law has been slightly modified on several occasions, the basic requirement still remains (Swedish Board of Agriculture, 2016): that cattle and sheep should be given the opportunity to graze during the summer period.

▶▶ Comparative studies of hay and silage in the 1970s

Total annual milk production remained about the same at three million tonnes from 1960 to 2000. However, the drastic decrease in the number of farms during the 1960s temporarily resulted in a smaller total amount of milk produced, which promoted intense research during the 1970s to increase production, in order to secure self-sufficiency of milk in the country. One of the main changes in cultivation during this time was that forage cultivation for winter feed moved from grazing areas to arable land with higher potential. This was necessary to keep winter feed production near the growing herd, and was possible because of increased crop yield brought about by mineral fertilisation.

In the 1970s, significant efforts were made at the Swedish University of Agricultural Sciences (SLU) to study the effects of conservation method and stage of silage/hay maturity when fed to dairy cows. Cuts at different stages of development and silage with or without wilting were also studied (Bertilsson, 1983). The studies concluded that silage produced more milk than hay. In one three-year experiment where 131 cows were given hay or silage in equal restricted amounts, the silage-fed cows produced 4% to 10% more milk. When harvest was postponed by 10 to 20 days, hay gave an 8% to 10% decrease in milk yield, while silage gave a minor decrease (Table SE1).

The silage system displayed many other benefits in practice. For example, it required a shorter period of dry weather and resulted in lower field losses than haymaking.

Since farm sizes had grown, the traditional hay racks were abolished in favour of field curing, which was easier to mechanise. However, when the dry hay was pressed in the field, many of the finer parts of clover and grass plants were over-dried and became brittle and were lost, resulting in a lower energy and protein concentration in the hay, although it was cut on the same day as the corresponding silage. These field losses were accounted for in an experiment where the hay was harvested at 60% dry matter (DM) and then dried further in the barn (Table SE1). However, in practice the hay was dried to higher DM levels when the weather was good.

The studies at SLU had a great impact on a dairy business in decline and in great need of methods to increase production.

Table SE1. Daily feed intake and animal performance data. Early cut at the boot stage and late cut 10 days later. Three-year ley fertilised with 89 kg N, species timothy, meadow fescue and red clover. Hay wilted to 60% in the field and to 87% DM in barn drying. Silage direct-cut, 26% DM. Forage fed restricted, concentrate according to milk yield. Lactation week 2-10, year 1 (From Bertilsson, 1983).

	Early cut hay	Early cut silage	Late cut silage
Forage intake, kg DM/day	8.7[a]	8.4[b]	8.5[b]
Concentrate intake, kg DM/day	8.1	8.4	8.6
Energy intake, MJ/day	197	199	194
Milk production, kg/day	26.1[a]	27.4[b]	26.2[a]
Milk fat content, %	4.56	4.65	4.57
Fat-corrected milk, kg ECM/day	28.6[a]	30.5[b]	28.3[a]

Means with different superscripts within rows are significantly different at $P < 0.05$.

▸▸ Conversion from hay to silage for dairy cows in the 1980s

The research findings from the 1970s and early 1980s were put into practical use through advisory organisations, facilitated by the fact that the number of dairy farms was decreasing rapidly and that the farmers were members of various cooperative advisory organisations. About 90% of dairy cows were included in the milk recording system organised by the Swedish Dairy Association, which also organised artificial insemination. Cows were test-milked once a month, the data were collected in a national database and most farmers had the feed ration recalculated to each individual cow after each test milking, based on the latest yield and the feeds available at the farm. This system required feed analysis of the roughage produced on the farm. Thanks to this system, farmers became aware that a higher nutrient content in their home-grown forage immediately resulted in a ration that contained less commercial concentrate, which meant less cash outflow from the farm.

Experiments had shown that the silage system resulted in lower field losses and higher nutrient concentration in the forage than the hay system when harvested at the same time. Although the difference was not enough as to be obvious in practice, the former tradition of harvesting hay at a late stage of development gave the new system an advantage. When silage making was introduced, it was made clear that it had to be harvested in the grass boot stage and it soon became obvious that silage always had a higher protein and energy content than hay.

▸▸ Decrease in roughage in the dairy cow diet in the 1990s and increase in the 2000s

In the early 1990s, the transition from hay to silage was almost complete in the dairy business. The production per cow increased with the shift to silage and in 1990 cows were producing 7,100 kg milk per cow per year, compared with 5,900 kg in 1980. After having achieved the positive response in milk production by changing to silage, the quest for larger production continued. The price ratio between concentrate and milk was such that it was profitable to increase concentrate to get more milk per cow. A summary of how high-producing cows were fed in Sweden from the mid-1970s until today gives an idea of how the feeding practice developed (Table SE2). In the example, the shift to silage and better forage quality was made between 1975 and 1982. Thereafter, the concentrate proportion increased steadily until some years into the new century. The reason why this was possible without negative consequences on animal health was the composition of the concentrates. The starch proportion was not allowed to increase to harmful levels and the cereals were successively replaced with fibre in the form of sugar beet pulp and other fibrous products. The amount of grain per cow remained at approximately the same level from 1986 and 2015. The trend towards larger concentrate amounts was broken some years ago due to concentrate becoming too expensive in comparison with high-quality forage. This is why silage is coming back in the dairy ration, even for high-producing cows. In 2017, annual milk production per Swedish cow in official milk reporting exceeded 10,000 kg energy-corrected milk (ECM).

Table SE2. Historical overview of winter feeding of high-yielding dairy cows in practice in Sweden.

Year	Feed ration	Milk yield per cow/year (kg)
1975	7 kg hay + 10 kg concentrate* (shift hay to silage)	5,500
1982	11 kg DM silage + 10 kg concentrate (75/10/15)*	6,200
1986	9 kg DM silage + 13 kg concentrate (60/20/20)*	6,700
1994	7 kg DM silage + 16 kg concentrate (50/30/20)*	7,600
2005	9 kg DM silage + 16 kg concentrate (50/30/20)*	8,500
2015	12 kg DM silage + 15 kg concentrate (50/30/20)*	10,000

* grain/sugar beet pulp/oil seed cake (%).

⯈ Shift to round bale silage among smallholders

Although silage was introduced as the forage conservation system on all dairy farms in the 1990s, a large proportion of grasslands was still harvested as hay. That was predominantly on smaller farms with beef cattle and, of course, as winter feed for the growing horse population. Beef in Sweden is mainly a by-product from the dairy herd. Male calves go to beef production and since the replacement percentage in the dairy herds is high, a great deal of beef also originates from cull cows of varying age. However, some production of specialist beef cattle takes place on small or medium-sized farms. The winter feeding season for this type of animal is shorter, lowering the quantity of conserved feed needed. It is always difficult to feed silage from a silo to few animals, since the low speed of use can easily lead to problems with heating of the silage. Farms handling few animals also have a low level of mechanisation and handling silage by hand is heavy work. However, turning to silage with larger nutrient content was attractive for beef farmers, who saw the possibility of raising the animals on forage only, without concentrate.

Farmers, always in the front line of development, tried their own way. Big round bale machines to harvest and store straw had become common among grain-producing farmers and were also used for hay in certain areas. The same farmers bought fertiliser in big plastic sacks. Some enterprising hay-producing farmers tried pressing wet grass instead of dry straw or hay in the bales, put them in the plastic fertiliser sacks and sealed them thoroughly – round bale silage was invented!

The idea spread and the round bale system soon became a subject for research institutes. The initial findings from most studies was that round bales were not well suited for silage making. Compared to silos, the surface area is much larger and contact with the air tends to destroy the outer surface of all silage. Therefore, the expectation was for great losses with this system as well as many problems with unwanted microbial activity such as enterobacteria, clostridia, yeast and mould. Such problems were found, but to a lower extent than expected and with great variability, which indicated that the system could be improved.

After several years with round bales in sacks, stretch film was introduced for wrapping and became a major success. The laborious work of putting the bales in sacks was removed and the elastic layers of film worked as an ideal one-way valve, letting gases escape from the bale but stopping air coming in. One of the major problems with the bales in sacks was that they blew up and were often punctured.

Trials soon reported good results with wrapped bales (Lingvall and Lindberg, 1989; Lingvall *et al.*, 1990, 1993) and the system spread rapidly. A survey of silage systems in use in 17 European countries found that forage wagon and metered-chop systems were most common in almost all countries, but Luxembourg, Norway, Sweden, Switzerland, Italy and the UK reported that big balers were increasing (Wilkinson and Starke, 1992). In Sweden, big balers were non-existent in the 1980 statistics, but were the five most sold in 1985 and the second most sold in 1990.

▸▸ Horse owners finally turn to silage

By the late 1990s, the only major category of grass-fed animals that were still on hay were horses. The horse population in Sweden has increased substantially in recent decades and horses now consume a great deal of the forage produced. Ensiling the grass was reported to reduce voluntary intake in horses, and horse owners sometimes claimed that silage was refused by their horses. Since no real comparison had been made between hay and silage of the same crop, cut at the same time and in the same field, an experiment was set up in which wrapped forage with 35% DM (silage) and with 55% DM or 70% DM (haylage) were compared with hay with 87% DM (Müller and Udén, 2007). When horses were presented with these feeds in a free choice system, silage was the first choice in 85% of observations and had the highest consumption rate. Hay had the lowest consumption rate and was never completely eaten. The haylages fell in intermediate order, with the drier types less popular. The conclusion was clear; horses like silage and haylage better than hay.

Other experiments dealt with different additives and DM levels (Müller, 2005). The silage was made using a conventional high-density hay baler that produced square bales with dimensions 80 cm × 48 cm × 36 cm. The idea was to produce silage that was easy to handle in horse stables and thus attractive to horse owners.

Recent studies comparing DM losses involved with different silo systems for silage resulted in surprising differences between the large silo structures and the wrapped bales. In field studies on farms, constructed silos (bunker and tower) had 14.1% DM losses and bag silos had 11.5% DM losses, while wrapped forage bales had DM losses of only around 1% (Spörndly, 2017). Given such results, it is not surprising that more forage is preserved in wrapped bales in Sweden (about 50%) than in bunker, bag or tower silos (Pettersson *et al.*, 2009).

▸▸ Grazing in Swedish production systems of today

Due to continuous investment by society in maintaining the biodiversity of semi-natural pastures, Sweden now has approximately 450,000 ha of semi-natural grasslands, many of them with high biodiversity values with regard to plants, insects, birds and other organisms. In addition to the permanent semi-natural grasslands, there are approximately 170,000 ha of grazed temporary grasslands in Sweden, as shown in Figure SE2.

Although most grazed temporary grasslands in Sweden consist of a mixture of species (e.g. meadow fescue, perennial ryegrass, smooth stalked meadow-grass and white clover), semi-natural pastures are even more heterogeneous. They consist of a mixture of dry, mesic and wet vegetation with a wide range of species. The proportions of these different types of vegetation vary between different pastures, depending on geographical location and soil conditions. The large difference in vegetation types also leads to a large variation in herbage production. Annual production of dry, mesic and wet vegetation in Swedish semi-natural pastures is reported to be approximately 1,800, 3,000 and 4,400 kg DM per ha, respectively

(Spörndly and Glimskär, 2018). In a similar manner, the forage nutritive value differs considerably, with the highest average annual content of metabolisable energy reported for mesic vegetation and the lowest for wet vegetation, according to weighted averages in recent studies (Spörndly and Glimskär, 2018). Furthermore, there are often considerable numbers of bushes in semi-natural pastures, often with thorns that animals will avoid, such as dog-rose and blackthorn. Trees are often larger, solitary trees such as birch and oak trees that provide shade for grazing animals.

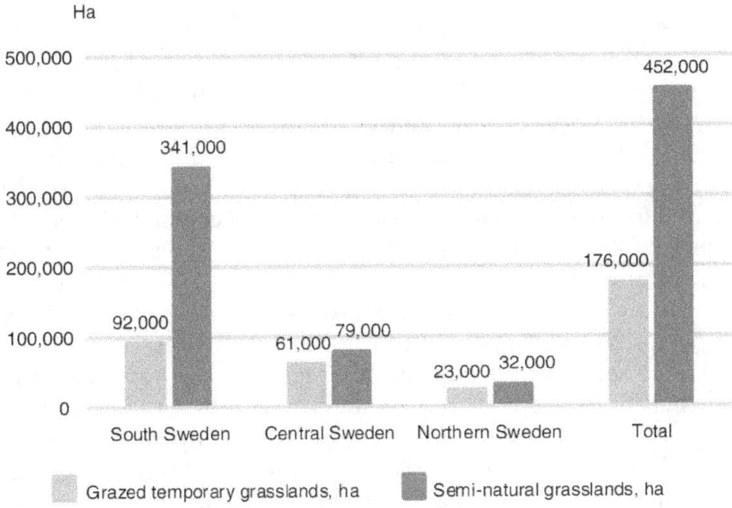

Figure SE2. Area of semi-natural grassland and grazed temporary grassland in Sweden, in 2017 (Swedish Board of Agriculture, 2017).

The semi-natural pastures are mainly grazed by cattle, with regard to acreage (Table SE3), but also with regard to the number of grazing sites and grazing animals according to a large inventory of a stratified sample of semi-natural grasslands (Spörndly and Glimskär, 2018). As shown in Table SE3, cattle dominated the pastures in the inventory, grazing approximately 65% of the acreage and a similar proportion of sites. Cattle breeds grazing these semi-natural pastures are approximately 50% dairy breeds (Swedish Red breed or Swedish Holstein) and 50% beef breeds or crosses with beef breeds. Horses and sheep each graze approximately 10% of the area of semi-natural pastures and mixed herds with several animal species are also common, especially on larger sites. Approximately 40% of the sites in the inventory were small, from 0 to 5 ha, but a few larger sites (5% >40 ha) constituted 40% of the total area studied (Spörndly and Glimskär, 2018). The large number of smaller sites is important with regard to maintaining biodiversity as they facilitate the spread of plant species between different areas, while the larger sites play an important role in providing a large area with similar and stable biological conditions.

Table SE3. Proportion of cattle, sheep and horses grazing semi-natural pastures in Sweden. N = 219 sites, with an average size of 14 ha. (Spörndly and Glimskär, 2018).

	% of acreage	% of grazing sites	% of grazing animals
Cattle	68	64	66
Horses	8	18	5.5
Sheep	9	11	28.5
Mixed[1]	15	7	

[1]Mainly cattle and sheep.

In contrast to semi-natural pastures, temporary grasslands are mainly grazed by cattle with higher nutrient demands, such as dairy cows. Under Swedish law, the required length of the grazing period depends on geographical location, ranging from 60 days in the north of Sweden to 120 days in the south. For all categories except lactating dairy cows, 24-hour grazing is required. For lactating dairy cows the daily grazing time is shorter, but they are required to come out to pasture daily and have access for at least six hours to a pasture area covered with vegetation that can be grazed simultaneously by all cows in the group. The amount of pasture that the animals are offered is not defined by law, however. With increasing herd size and approximately one-third of all milk produced in automatic milking systems, the use of pasture as part of the diet has decreased substantially. Thus, many farmers provide their animals with a smaller grazing area with some opportunity to graze, primarily for welfare and recreation to fulfil the legal requirements. This system is often referred to as exercise pasture. Cows with access to exercise pasture are still offered full indoor feeding with concentrates and silage during most of the summer, to ensure that their nutrient needs are covered at all times. Swedish consumers have a keen interest in animal welfare and, in 2018, this led the largest dairy to introduce a price premium to farmers who allow their cows to have 25% longer access to pasture (i.e. 7.5 hours) than the stipulated minimum of six hours required by law (Arla, 2018). Together with the comparatively low cost of pasture, this may create increased interest in pasture as a feed for dairy cows in coming years.

▸▸ Organic production in Sweden

In Sweden, organic production is increasing in all sectors. This is particularly pronounced in forage and cattle production. Table SE4 summarises the extent of organic production in 2017. In organic dairy production, the main organisation that certifies organic milk (KRAV, 2018) stipulates that pasture intake must be at least 6 kg DM daily. Furthermore, dairy cows must be kept on pasture for more than 12 hours daily, a major difference compared to conventional production.

Table SE4. Organic production in proportion (%) of total agricultural land or animal stock (Swedish Board of Agriculture, 2018a).

Organic production	Proportion of total agricultural area or animal stock (%)
Total acreage	19.1
Cereal grain (wheat, barley, oats, rye, triticale)	9.5
Forage (legumes/grass, green fodder, ploughed)	22.1
Semi-natural pasture	24.6
Dairy cows	16.4
Beef cows	33.7
Sheep	20.9
Pigs	2.3
Laying hens	17.0
Broilers	1.9

▶▶ How the new conditions influenced seed selection

Forage species and varieties well adapted to different purposes and regions are crucial for high quality and quantity in forage production and thus beef/dairy farm profitability. The problem with introducing winter-hardy plant material, such as from Canada into Sweden, is that growth starts earlier in spring and stops later in autumn in Sweden than in Canada. It is necessary to breed and test plant material specifically for Sweden, where SLU runs the official variety testing programme (VCU) (Halling and Larsson, 2017). The plant material mainly consists of Swedish-bred varieties. Lantmännen Lantbruk has active breeders of forage crops such as timothy, meadow fescue, perennial ryegrass, cocksfoot, lucerne, red clover and white clover, but a large proportion of the forage seed sold comes from other countries validated for the EU list. Fortunately, most of these varieties are also validated for Swedish conditions.

Characteristics that are important to take into account with regards to the local seasonal growing pattern, competitiveness, resistance to pests and winter hardiness of species/varieties include the role in cropping, grazing and feeding system; persistence; cutting regime; and fertilisation regime.

Since 2002, the amount of sown legumes in Swedish temporary grasslands has increased from 16.3% to 18.6% of the total amount of certified and imported seed of the main forage species for silage production (Swedish Board of Agriculture, 2018b) (Table SE5). The ratio of different species has changed over time, with red clover having decreased and white clover and lucerne having increased. In recent years, there have been studies on fibre quality in timothy, leading to improved varieties and more sown timothy than before. There have also been some setbacks in the use of ryegrasses, as despite the nutritional advantages of these high-yielding grasses, the climate conditions in Sweden sometimes prevent efficient use of these species.

Accordingly, timothy, the most winter-hardy forage grass, has received new attention. Following successful breeding results in recent years, the use of tall fescue has also increased. Tall fescue is a good example of a species for which breeders have recently improved the quality and palatability in combination with existing good performance (Halling and Larsson, 2017). Good tolerance to both wet and dry conditions is a desirable characteristic in a changing climate.

Table SE5. Certified and imported seed of the main silage crops in Sweden (10^3 kg/year). Percentage forage legumes/forage legumes and grasses (Swedish Board of Agriculture, 2018b).

Species		2002/2003		2009/2010		2017/2018	
English name	**Latin name**	**10^3 kg**	**%**	**10^3 kg**	**%**	**10^3 kg**	**%**
Red clover	*Trifolium pratense* (L.)	806.7	78	550.4	64	473.8	42
White clover	*Trifolium repens* (L.)	155.8	15	139.9	16	430.5	38
Alsike clover	*Trifolium hybridum* (L.)	13.8	1	31.5	4	40.5	4
Lucerne	*Medicago sativa* (L.)	60.0	6	127.5	15	187.3	16
Birdsfoot trefoil	*Lotus corniculatus* (L.)	4.2	0	6.7	1	9.1	1
Total forage legumes		*1 040.5*	*100*	*856*	*100*	*1 141.2*	*100*
Timothy	*Phleum pratense* (L.)	2 072.9	39	1 859.9	46	2 177.3	44
Meadow fescue	*Festuca pratensis* (Huds.)	1 567.9	29	722.9	18	1 502.3	30
Tall fescue	*Festuca arundinacea* (Schreber)	29.0	1	253.1	6	312.4	6
Perennial ryegrass*	*Lolium perenne* (L.)	1 513.5	28	1 022.3	25	777.4	16
Hybrid ryegrass	*Lolium x boucheanum* (Kunth)	40.5	1	-	-	29.0	1
Festulolium *ssp.*	*x Festulolium*	49.0	1	96.1	2	-	-
Cocksfoot	*Dactylis glomerata* (L.)	69.8	1	68.6	2	196.0	4
Total forage grasses		*5 342.6*	*100*	*4 022.9*	*100*	*4 994.6*	*100*
Legumes/legumes + grasses			16.3		17.5		18.6

* Includes amenity varieties.

To conclude, the main breeding targets for forage crops in Sweden are large yield, persistence (i.e. winter hardiness and pest resistance), high nutritive value (e.g. digestibility, crude protein and fibre quality) and satisfactory seed production.

▸▸ More intensive harvesting regimes

Traditionally, cutting frequency was restricted to two cuts a year in Sweden due to the short growing season. Several investigations have shown that production is greater, but quality is lower, with two cuts compared with three cuts a year in mixed swards with timothy, meadow fescue, and in some treatments red clover, a commonly used seed mixture (Table SE6).

Table SE6. Dry matter yield (kg DM/ha) and digestible energy (MJ/kg DM) in grass and mixed swards with two and three cuts a year, with 100 kg N/ha supplied in both cases (Kornher, 1982).

Seed mixture	2 cuts		3 cuts	
	Yield	Energy	Yield	Energy
Timothy + meadow fescue	7,760	10.0	6,210	10.9
Timothy + meadow fescue + red clover	9,110	10.0	8,035	10.3

▸▸ White clover for silage

To improve forage-based diets in terms of required nutritional quality and quantity, the choice of species and of suitable management strategies is crucial. High digestibility, ensuring large forage intake and slow decline with time are major advantages in legumes compared to grasses, especially when the weather is unpredictable. Traditionally, white clover was grown for grazing in Sweden. However, in the 1980s, more erect varieties of white clover were introduced and studies were carried out to test the potential for inclusion of white clover in mixed, short-term leys.

An extensive study (15 field trials) was carried out with two different mixtures of white or red clover with timothy and meadow fescue, and smooth-stalked meadow grass (*Poa pratensis* L.). Three nitrogen levels were included (0, 100 and 200 kg N/ha). The swards were cut three (silage or hay developmental stages) or four times a year for four consecutive years (Svanäng and Frankow-Lindberg, 1994). On average for different fertilising and cutting regimes, the white clover/grass mixtures (WC) were found to be better than the red clover/grass (RC). The yield from WC with no fertiliser N supplied was about the same in the fourth year as in the first year (7,700 kg DM/ha) (Table SE7). The corresponding yield in RC decreased from 8,400 to 5,500 kg DM/ha, resulting in an increasing amount of weeds. Irrespective of harvesting regime, root rot is the major obstacle to more long-lived RC swards. An early first cut and short defoliation intervals increased the WC content. Nitrogen fertilisation increased the DM yield, but decreased the clover percentage and the clover yield in the sward. Due to less competitiveness, WC content was more depressed than RC content when N was applied. The marginal effect of N on DM yield was largest in the 0-100 kg N/ha interval (Table SE8). It was larger in RC than in WC except in the first year with 100 kg N/ha. The introduction of white clover facilitated more flexible sward management, sometimes with harvesting and grazing in combination.

Table SE7. Effect of nitrogen application on dry matter yield (kg DM/ha) in mixed swards fertilised with 0, 100 and 200 kg N/ha. Mean for *Trifolium* ssp. and harvesting systems (Svanäng and Frankow-Lindberg, 1994).

N level	*Trifolium* ssp.	1st year	2nd year	3rd year	4th year
0 kg N	Red clover	8,430	7,647	6,363	5,491
	White clover	7,724	8,085	7,441	7,765
100 kg N	Red clover	9,597	9,359	7,795	7,141
	White clover	9,123	9,534	8,169	8,152
200 kg N	Red clover	10,409	10,202	8,569	8,213
	White clover	9,920	10,246	8,619	8,565

Table SE8. Dry matter yield (kg DM/kg N) in mixed swards on the margin of nitrogen application in the intervals 0-100 kg N/ha and 100-200 kg N/ha. Mean for *Trifolium* ssp. and harvesting systems (Svanäng and Frankow-Lindberg, 1994).

N level	*Trifolium* ssp.	1st year	2nd year	3rd year	4th year
0-100 kg N	Red clover	11.8	17.4	14.5	13.0
	White clover	14.4	14.7	7.7	7.2
100-200 kg N	Red clover	8.3	8.7	8.2	10.1
	White clover	8.2	7.3	4.8	3.9

Most of the available forage seed mixtures contain both red and white clover. Frankow-Lindberg *et al.* (2009) found that red clover as a single legume species or in a mixture was superior at a dry site, while multi-clover/grass species mixtures were superior at a wet site. Stability of clover yields can generally be increased by including both white and red clover in the seed mixture, but not total DM yield.

▸▸ Using white clover for less intensive silage systems

White clover is more persistent than red clover and higher yielding in intensively cut systems. The studies in the late 1980s led to growing interest among Swedish farmers in using white clover as a silage crop. The question was whether white clover could be recommended in areas with less intense production systems due to farming tradition and climate conditions.

In a study with 11 field experiments in southern and central Sweden, mixed swards containing red or white clover were cut two or three times a year for three years following establishment, and fertilised with 0 or 100 kg N/ha (Nilsdotter-Linde *et al.*, 2002). The number of cuts significantly affected DM yield, but the response varied between sites and with sward age (Figure SE3). The effect of N on DM yield was positive, on average, in both red and white clover/grass mixtures. Nitrogen fertilisation rate significantly ($P < 0.001$) affected DM yield of individual cuts and total yield per year (Figure SE3). The number of cuts also affected yield, with two cuts giving larger total yields than three, an effect that increased with sward age. On average, the difference between red and white clover yield was small, but in the third year unfertilised white clover yielded more than unfertilised red clover.

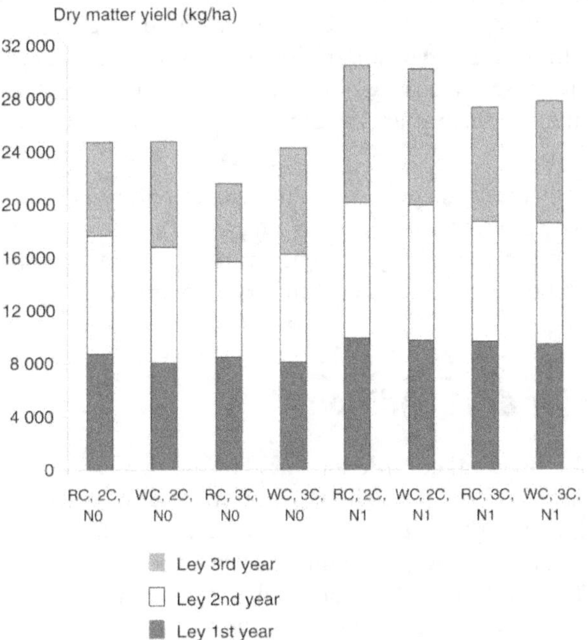

Figure SE3. Total dry matter yield for combined treatments (all cuts and years) as an average of 11 field experiments. RC = red clover, WC = white clover, 2C = 2 cuts/year, 3C = three cuts/year, N0 = 0 kg N/ha, N1 = 100 kg N/ha (Nilsdotter-Linde *et al.*, 2002).

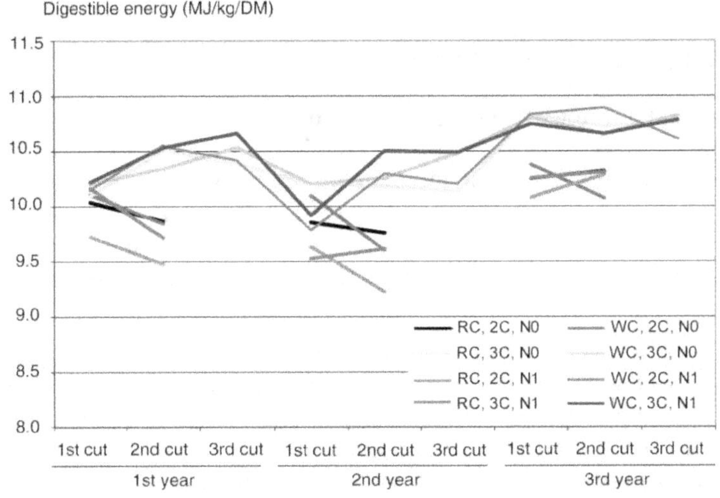

Figure SE4. Digestible energy calculated from rumen-soluble organic matter as an average of nine field experiments in all treatments at each cut in years I-III. RC = red clover, WC = white clover, 2C = two cuts/year, 3C = three cuts/year, N0 = 0 kg N/ha, N1 = 100 kg N/ha (Nilsdotter-Linde *et al.*, 2002).

The number of cuts had the largest effect on the mean nutritional quality of the herbage, especially digestible energy (MJ/kg DM) (Figure SE4), although at quite low levels in many cases (<10.5 MJ/kg DM). Neutral detergent fibre was higher in N-fertilised treatments than with no fertiliser N, while crude protein was higher in treatments with three cuts and where the legume content was high. That was the case in plots with no fertiliser N supplied and where the white clover fraction increased, as occurred during the third year. In some cases, the content of crude protein was above 200 g/kg DM.

The conclusion was that total yields of white clover and red clover in mixed swards were similar in young swards and with two cuts per year. However, unfertilised white clover yielded more in the third year. The nutritional quality, especially digestible energy, was much better with three cuts than with two.

▶▶ Lucerne for dry conditions

In the early 1980s, there was considerable field research on lucerne in Sweden, which was soon implemented by farmers in appropriate regions with a high soil pH. Soil inoculation with a suitable *Rhizobium* strain improves lucerne yield substantially (Jönsson, 1982). Following successful introduction, there were some setbacks owing to hard winter damage followed by very sparse swards. The winter buds just below the soil surface need oxygen and are susceptible to standing water and ice coverage. Field topography and drainage status are very important when including lucerne in leys. In recent years, there has been renewed interest in this crop for reasons such as the high fibre quality, the need for more home-grown protein in the ration and perhaps more pronounced dry periods in the summer.

Ryegrasses: possibilities and threats

With the arrival of intensive harvesting systems with more white clover included for silage, ryegrasses have become more interesting, even at Swedish latitudes. Large yield in combination with high quality is their major advantage. However, the reason why these species are not as prevalent in Scandinavia as they are further south is that their longevity is more or less restricted depending on climate conditions, with winter kill always a threat. Perennial ryegrass is only recommended in the southern third of Sweden.

Swedish farmers have very good knowledge about management of domestic timothy, but are less knowledgeable about different treatments for improving the overwintering capacity of ryegrasses. In nine field experiments with perennial ryegrass in the official testing programme in eastern and western Sweden in the early 1990s, late autumn cutting was tested as a tool to reduce damage caused by such issues as snow mould (*Fusarium nivale*) and thus improve winter survival (Halling, 1994). Because of extremely mild winters with low occurrence of snow mould fungi in the study period, cutting as late as possible before cessation of growth significantly reduced the following spring yield by about 25% in both the second and third year

(Table SE9). There was no residual effect of late autumn cutting on subsequent cuts.

Table SE9. Effect of autumn cutting management on subsequent dry matter (DM) yields in perennial ryegrass (10^3 kg DM/ha, mean different varieties) (Halling, 1994).

Treatment	DM yield				3rd year
	2nd year				
	1st cut	2nd cut	3rd cut	Total yield	1st cut
With late autumn cut	3.31	1.94	2.56	7.81	2.40
Without late autumn cut	4.46	1.92	2.71	9.11	3.25
Significance	**	NS	NS	*	**

Significance: NS = $P > 0.05$, *$P < 0.05$, **$P < 0.01$, ***$P < 0.001$.

The growing interest in different ryegrasses and their hybrids is encouraging breeders and researchers to focus on better varieties and management strategies, and there are ongoing investigations on how cutting strategy influences overwintering capacity.

▸▸ Forage species adapted for different purposes

There is increasing interest in species with special qualities, e.g. water-soluble carbohydrates and condensed tannins. Studies on birdsfoot trefoil, a minor forage legume containing condensed tannins, have confirmed that it can withstand Swedish climate conditions. This has led to an extensive trans-disciplinary investigation of its population ecology, protein efficiency and anti-parasitic effects in ruminants (Nilsdotter-Linde *et al.*, 2002), followed by a performance investigation in heifers (Nilsdotter-Linde *et al.*, 2004) and a corresponding investigation in dairy cows. A tendency for larger milk yield and somewhat higher milk protein concentration resulted in higher protein yield with a birdsfoot trefoil diet compared with a white clover diet (Eriksson *et al.*, 2012). The most appropriate variety for Swedish conditions is cv. Oberhaunstaedter, which has shown good persistence and relatively high content of condensed tannins (1-2 g/kg DM) in the official variety testing programme (Halling and Larsson, 2017).

▸▸ Conclusions and future tasks

In Sweden, ruminants and horses rely on preserved forage, as the grazing season is about four months in the south and only two months in the north. In the 1960s, haymaking dominated completely, but over the following 20 years there was a dramatic change to silage making. Less dependency on good harvesting weather, better technical solutions and better animal response were the most important factors for this change. The large dairy herds were the first to change to grass and legume silage, stored in tower or bunker silos. When round bales arrived in the 1990s, smallholders too began making silage, while horse owners converted when small bales and haylage were introduced.

When silage replaced hay, the cropping system changed to more intensively managed, short-term leys. Two cuts were replaced by three in central and northern Sweden and four or more in southern Sweden. Sward composition also changed, with white clover included for silage and with ryegrasses becoming more popular. However, ryegrass still has problems with winter kill and remains a minority grass in Swedish leys. It is a challenge to find the right plant material combining persistence, large yield and high quality. Farmers have to formulate their own strategy regarding longevity, fertilisation, harvesting intensity and the purpose of production in choosing the right seed mixture.

There is currently renewed interest in producing home-grown protein to replace imported protein. Temporary grasslands provide a large proportion of the protein needed, but more can be done to produce large yields of the proper quality for different animal categories. More emphasis has to be put on plant breeding and improved management to achieve more persistent swards for grazing and harvesting. According to recent research and experiences from other countries, improved grazing management can make a major difference in the economics of grazing. Deeper knowledge about cultivation of grasses and legumes that are more tolerant to dry and/or wet conditions is important, to better face upcoming climate changes.

Within research on silage production from grasslands, DM losses during silage making are gaining interest. The magnitude of DM losses differs significantly between the silo systems in use. Our research showed that the losses in tower silos and bunker silos were about 15%, while those in round bale silos were around 1%. These figures indicate that, with maintained production level, the total acreage cultivated as forage could be decreased by about 10% through appropriate management of silos and silage making processes.

▸▸ References

Arla, 2018. Arla launches Summermilk® products with longer time on pasture during the summer. In Swedish. Accessed Jan 15, 2019. http://www.mynewsdesk.com/se/arla/pressreleases/arla-lanserar-sommarmjoelk-r-produkter-och-laengre-sommarbete-2481223

Bertilsson J., 1983. Effects of conservation method and stage of maturity upon the feeding value of forages to dairy cows. Dissertation. Swedish University of Agricultural Sciences. Department of Animal Husbandry. Report 104. Uppsala, Sweden.

Eriksson T., Norell L., Nilsdotter-Linde N., 2012. Nitrogen metabolism and milk production in dairy cows fed semi-restricted amounts of ryegrass-legume silage with birdsfoot trefoil (*Lotus corniculatus* L.) or white clover (*Trifolium repens* L.). *Grass and Forage Science*, 67 (4), 546-558.

Frankow-Lindberg B.E., Halling M., Höglind M., Forkman J., 2009. Yield and stability of yield of single- and multi-clover grass-clover swards in two contrasting temperate environments. *Grass and Forage Science*, 64 (3), 236-245.

Halling M.A., 1994. Effect of autumn treatment on winter survival of cultivars of perennial ryegrass (*Lolium perenne*) under Swedish conditions. 8th General Proceedings of the 15th General Meeting of the European Grassland Federation. Wageningen, The Netherlands, pp. 177-180.

Halling M.A., Larsson S., 2017. Forage species for cutting, grazing and green fodder. Varieties for south, central and northern Sweden 2017/2018. Swedish University of Agricultural Science. Department of Crop Production Ecology. 77 pp. https://www.ffe.slu.se/Info/sortval_2017-2018.pdf

Jönsson N., 1982. Blålusern. Resultat av odlingstekniska försök. Sveriges lantbruksuniversitet. Institutionen för växtodling. Rapport 99. 33 pp. In Swedish.

Kornher A., 1982. Vallskördens storlek och kvalitet. Inverkan av valltyp, skördetid och kvävegödsling. Sveriges lantbruksuniversitet. *Grass and Forage Reports,* 1, 5-32. In Swedish.

KRAV, 2018. Regler för KRAV-certifierad produktion utgåva 2018. KRAV ekonomisk förening. Uppsala, Sweden. 308 pp. In Swedish.

Lingvall P., Lindberg H., 1989. High quality silages by wrapping big bales. Big bale silage conference. British Grassland Society, pp. 5-10.

Lingvall P., Lindberg H., Jonsson A., 1990. The impact of packing technique, plastic film and silage additives on round bale silage quality. Proc. 13th General Meeting of the European Grassland Federation. Banska Bystrica, Czechoslovakia, pp 240-248.

Lingvall P., Pettersson C.M., Wilhelmsson P., 1993. Influence of oxygen leakage through stretch film on quality of round-bale silage. Proc. XVII International Grassland Congress, New Zealand.

Müller C.E., 2005. Fermentation patterns of small-bale silage and haylage produced as a feed for horses. *Grass and Forage Science*, 60, 109-118.

Müller C.E., Udén P., 2007. Preference of horses for grass conserved as hay, haylage or silage. *Animal Feed Science and Technology*, 132, 66-78.

Nilsdotter-Linde N., Stenberg M., Tuvesson M., 2002. Nutritional quality and yield of white or red clover mixed swards with two or three cuttings with and without nitrogen. *Grassland Science in Europe*, 7, 146-147.

Nilsdotter-Linde N., Olsson I., Hedqvist H., Jansson J., Danielsson G., Christensson D., 2004. Performance of heifers offered herbage with birdsfoot trefoil (*Lotus corniculatus* L.) or white clover (*Trifolium repens* L.). *Grassland Science in Europe*, 9, 1062-1064.

Nilsdotter-Linde N., Bernes G., Christensson D., Hedqvist H., Jansson J., Murphy M., Tuvesson M., Waller P., 2002. Birdsfoot trefoil (*Lotus corniculatus* L.) with respect to agronomy, *in vitro* protein degradability and parasitic infections in grazing animals. Proceedings of the 14th IFOAM Organic World Congress. Victoria, British Colombia, Canada. 21-24 August, pp 92.

Pettersson O., Sundberg M., Westlin H., 2009. Machinery and methods in forage production. Institutet för jordbruks- och miljöteknik. Uppsala, Sweden. Report 377. In Swedish.

Spörndly E., Glimskär A., 2018. Grazing animals and grazing pressure in Swedish semi-natural pastures. Swedish University of Agricultural Sciences. Department of Animal Nutrition and Management Report 297, 71 pp. In Swedish.

Spörndly R., Nilsdotter-Linde N., 2011. L'ensilage des prairies temporaires en Suède: Un développment réussi. *Fourrages* (Revue de l'Association Francaise pour la Production Fourragère), 206, 107-117.

Spörndly R., 2018. Dry matter losses from different silo structures. Proceedings of the 9th Nordic Feed Science Conference. Uppsala, Sweden, pp. 171-176.

Swedish Board of Agriculture, 2016. SJVFS, 2016:13 L100:6. Regulations of changes in the animal welfare ordinance SJVFS 2010:15, 1-6.

Swedish Board of Agriculture, 2017. The yearbook of agricultural statistics 2017. Swedish Board of Agriculture and Statistics Sweden (SCB).

Swedish Board of Agriculture, 2018a. The yearbook of agricultural statistics 2018. Swedish Board of Agriculture and Statistics Sweden (SCB).

Swedish Board of Agriculture, 2018b. Certifiering 2017/2018. http://www.jordbruksverket.se/download/18.25e61c93165e6d49047c6368/1537450549546/%C3%85r%202017-2018.

Svanäng K., Frankow-Lindberg B., 1994. White clover leys. Effect of nitrogen fertilisation and harvest systems. Sveriges lantbruksuniversitet. Institutionen för växtodlingslära. Växtodling 51. 23 pp. With English summary.

Wilkinson J.M., Starke B.A., 1992. Silage in Western Europe. 2nd edition. Chalcombe publications. Southampton, UK.

Ireland

Fergus Bogue and Michael O'Donovan

▸▸ Introduction

Grazed grass is the cheapest and most widespread feed for ruminant production systems in Ireland. Grass enables low-cost animal production and promotes a sustainable, green and high-quality image of milk production across the world. Recent industry reports (FoodHarvest 2020 and FoodWise 2025) have highlighted the important role grass can play in an expanding milk production industry. Through a combination of climate and soil type, Ireland possesses the ability to grow large quantities of high quality grass and convert it through the grazing animals into high-quality grass-based milk and meat products.

Our competitive advantage in milk production can be explained by the relative cost of grass, silage and concentrate feeds. Therefore, increased focus on grass production and efficient utilisation of that grass should be the main driver for expansion of the livestock sector. An analysis of farms completing both grassland measurement in PastureBase Ireland and a Profit Monitor demonstrated increased profit of €181/ha for every one tonne DM/ha increase in grass utilised. It should be noted that issues such as environmental sustainability (carbon footprint, nutrient use efficiency etc.) are also improved by increased grass utilisation.

Future growth in pasture-based milk production in Ireland will depend on an effective grass-based system. However, Irish farmers are not using grass to best effect and there is thus a need to (1) increase grass production and (2) ensure efficient utilisation of that grass.

▸▸ Current grazing performance on dairy farms

Currently, it is estimated that about eight tonnes grass DM/ha is utilised nationally on dairy farms (Dillon, 2016). There are major improvements required in areas of pasture production and utilisation. Data from the best commercial grassland farms and research farms indicate that the current level of grass utilised can be increased significantly on dairy farms (greater than 10 t DM/ha utilised – i.e. 14 tons DM/ha grown and 75% utilisation rate).

It is important to recognise that improvements in soil fertility, grazing infrastructure and the level of reseeding are in achieving higher levels of grass production and utilisation. However, to achieve greater change in the level of grass utilised, farmers will need to upskill their grazing management practices. This means regular measurement of grass cover, using specialised grassland focused software to analyse grass production and making and implementing grazing management decisions. These are key drivers to increasing grass production on the farm. New technologies are now available which make grass cover assessment and the decision-making process much easier.

▸▸ Soil fertility management

Good productive soils are the foundation of any successful farming system and key for growing sufficient high-quality grass to feed the herd. The management of soil fertility levels should therefore be a primary objective of every farm. A recent review of soils tested at Teagasc indicates that the majority of soils in Ireland are below the target levels for pH (i.e. 6.3) or phosphorus (P) and potassium (K) (i.e. Index 3) and will be very responsive to application of lime, P and K. On many farms suboptimal soil fertility will lead to a drop in output and income if allowed to continue. Teagasc highlights five steps for effective soil fertility management:

1. Have soil analysis results for the whole farm (soil sampling every 2 years)

2. Apply lime as required to increase soil pH up to target pH for the crop

3. Aim to have soil test P and K in the target Index 3 in all fields

4. Use organic fertilisers as efficiently as possible

5. Make sure the fertilisers used are properly balanced

For those farmers aiming to improve soil fertility on their farms, following these five steps provides a solid basis for success.

Phosphorus (P)

The proportion of soils tested with low soil P fertility (i.e. P Index 1 and 2) has increased to approximately 62%. This overall trend reflects the soil P fertility status on many farms, and indicates a serious loss in potential productivity. Recent research has shown that soils with P Index 3 will grow approximately 1.5 t (DM)/ha per year more grass than soils with P Index 1. Most of the DM yield response in these experiments took place in spring and early summer.

Potassium (K)

Soil analysis shows that the trend in soil K status, across dairy and drystock enterprises, broadly mirrors that for P. Despite no legislative limits on K fertilisers, K usage dropped in line with P fertiliser applications. Consequently, soil test results indicate a sharp increase in soils with low K status. Over half of the soil samples tested by Teagasc had very low to low soil K status (i.e. K Index 1 or 2).

Increasing soil nutrient availability: lime

Lime is a soil conditioner and corrects soil acidity by neutralising the acids present and allowing the microorganisms and earthworms to thrive and break down plant residues, animal manure and organic matter. This helps to release stored soil nutrients such as nitrogen (N), P, K, sulphur (S) and micro-nutrients for plant uptake. In addition, ryegrass and clover swards will persist for longer after reseeding where soil pH has been maintained close to the target levels through regular lime applications.

Liming acidic soils to correct soil pH will result in the following:
- Increased grass and crop production annually
- Increase the release of soil N by up to 60 units N/acre/year
- Increase the availability of soil P and K and micronutrients
- Increase the response to freshly applied N, P and K as either manures or fertiliser

Ground limestone is the most cost effective source of lime and can be applied throughout the year when the opportunity arises. Lime is the foundation of soil fertility and is a primary step to take when correcting soil fertility.

Investing in grazing

In order for expansion to be successful, many farms will need to make significant investments. The available capital for this investment will be scarce as expansion happens and continues. Therefore, investment on the farm should be prioritised in areas that increase efficiency and reduce the exposure of the business to external shocks such as lower prices of products or higher prices of inputs etc. All investments that give the highest returns should be prioritised.

Every ton of additional grass eaten by grazing animals will add €180/ha additional profit to a dairy farm. Therefore it is important for investment in grazing to be prioritised to give the maximum return. The table IE1 summarises the potential return on investment for different investments in a dairy farm business. The level of return on these investments is high because it means investing in grazing. These investments will enable the farm to grow more grass, lengthen the grazing season or both.

Table IE1.Potential return on investment for different investments in a dairy farm business.

Investment	Cost	Impact	Annual return (%)
Increase soil P and K levels	P and K application of 20 and 50 kg/ha	+1.5 t DM/ha/year grass growth	152
Reseed full farm in eight year cycle	€650/ha	+1.5 t DM/ha/year grass growth	96
Improve grazing infrastructure	€1,000/ha for roads, fencing and water	+1.0 t DM/ha/year grass growth	58

The need for more reseeding

Because grass is our primary feed during the main grazing season, and the primary source of winter forage in the form of grass silage, the low level of reseeding must be addressed. Reseeding must be combined with managing, and where necessary, increasing soil fertility. Ireland will continue to increase milk production and the focus on efficient production of this milk is critical to maintain our industry competitiveness. Teagasc have developed a national grassland database (PastureBase Ireland), and the initial results show that there is huge capacity on Irish farms to grow more grass. The objective here is to outline the key points in grassland reseeding and to ensure farmers making the investment in renovating grassland get the best possible results.

Why reseed

Productive grassland farms must have perennial ryegrass dominated swards. Recent Moorepark research shows that old permanent pasture produces, on average, 3 t DM/ha per year less than perennial ryegrass dominated swards. Old permanent pasture is up to 25% less responsive to available nutrients such as nitrogen than a perennial ryegrass dominated sward. Reseeding is a highly cost effective investment. With regular reseeding the grass growth capacity of the farm can be increased substantially and the annual return of investment is large.

Objectives of reseeding are to create swards that:
– Increase the overall productivity of the farm
– Increase grass quality
– Are responsive to fertiliser – at least 10 kg DM/kg N applied
– Allow higher animal output – 8% higher milk output per hectare relative to permanent pasture
– Increase grass utilisation
– Reduce silage requirement
– Increase the productivity of the farm (carry a higher stocking rate)
– Can allow clover to establish

Reseeding checklist
– Identify paddocks for reseeding (poorer performing paddocks; low perennial ryegrass content)
– Soil test and lime
– Sowing date
– Method of reseeding
– Spray off paddock
– When cultivating – prepare a good seed bed
– Choose appropriate grass cultivars
– Sowing rate
– Roll
– Slug and other pests
– Control weeds early
– Graze at two-leaf stage
– Avoid poaching and overgrazing

Cultivation techniques

How paddocks are prepared for reseeding depends on soil type, the amount of underlying stone and machine/contractor availability. There are many different cultivation and sowing methods available. All methods, when completed correctly, are equally effective.

Key points

– Spray off old sward
– Graze sward tightly or mow to minimise surface trash
– Apply lime
– Choose a method that suits your farm
– Soil test
– Firm fine seedbed with good seed/soil contact is essential
– Roll after sowing

Table IE2.

	Do's	Do not's
Ploughing	Shallow plough. Develop a fine, firm and level seedbed	Plough too deep (> 15 cm). Cloddy, loose seedbed
Discing	Graze light, apply lime. 3-4 runs in angled directions	Forward speed too fast – rough, uneven seedbed
One-pass	Graze tight, apply lime. Slow fossoirs speed at cultivation	Forward speed too fast – rough, patchy seedbed
Direct drill	Graze tight, apply lime and slang pellets. Wait for moist ground conditions (slight cut in ground)	'Trashy' seedbed – no seed/soil contact. Use when ground is dry and hard

Variety choice in perennial ryegrass

The Irish Department of Agriculture, Food and the Marine (DAFM) publishes the Recommended List, which shows the Pasture Profit Index values and agronomic values of the evaluation on the same table (see https://www.teagasc.ie/crops/grassland/pasture-profit-index/).

The Recommended List has evaluated varieties across years and sites and is the only evidence available of the potential performance of grass cultivars in Ireland. Using varieties not on this list basically reflects poor decision-making, as is buying grass seed on price. The varieties you use on the farm will be there for eight to 12 years, and choosing to use cheap mixes with non-recommended varieties will increase the chances of those varieties failing to perform on the farm.

Pasture Profit Index (PPI)

	Sub-indices (€ per ha per year)				Total
Variety	Seasonal Yield	Quality	Persistency	Silage	Merit
AberClyde	142	55	0	28	225
AberMagic	170	30	0	17	217
Fintona	152	24	0	39	215
AberZeus	169	9	0	34	212
Nifty	194	-12	0	26	208
Moira	198	-18	0	27	207
AberGreen	186	16	0	4	206
AberPlentiful	154	29	0	20	203
AberGain	119	60	-11	30	198
AberChoice	122	59	0	13	194
Meiduno	146	27	0	21	194
Dunluce	125	37	0	30	192
Elysium	138	31	0	20	189
AberWolf	142	25	0	21	188
Seagoe	129	14	0	42	185
Astonconqueror	149	7	0	24	180
AberBite	102	51	-11	33	175

	Sub-indices (€ per ha per year)				Total
Variety	Seasonal Yield	Quality	Persistency	Silage	Merit
Rosetta	158	-4	0	20	174
Solas	122	26	0	19	167
Kintyre	116	27	-5	18	156
Astonenergy	86	59	0	8	153
Xenon	91	40	0	19	150
Carraig	121	-16	0	35	140
Solomon	140	-28	0	25	137
Alfonso	82	45	0	7	134
Aspect	84	33	0	14	131
Navan	96	12	0	16	124
Drumbo	93	39	-11	0	121
AberLee	76	42	0	3	121
Kerry	113	-5	0	11	119
Glenroyal	104	-3	0	11	112
Clanrye	90	-15	0	20	97
Majestic	109	-26	0	5	88
Glenveagh	70	-21	0	12	61

Figure IE1. Pasture Profit Index (PPI).

When the decision to reseed is made, the next major decision is to select the most appropriate grass variety or varieties. The first thing to consider is the primary target use of the field. Will it predominantly be used for grazing or will it generally be used as a silage paddock? How much tetraploid should be used? A balance between quality, dry matter productivity and sward density is generally what must be achieved.

The key traits in a seasonal grass-based production system are:
− High quality
− High seasonal production
− Good persistency score

Table IE3. Differences between diploid and tetraploid perennial ryegrass varieties.

Tetraploid varieties	Diploid varieties
Tall upright growth habit	Prostrate growth habit
Create more 'open' sward	Create a denser sward with less 'open' spaces
Higher digestibility value	Generally lower digestibility and yield

Combining diploids and tetraploids in a mixture will create a dense, high quality sward – ensure you select varieties which express high performance in the key traits. Increasing the proportion of diploids on heavier soils is recommended to create better ground cover, however tetraploids should be used on heavy soils. Choosing all dense varieties will compromise DM production and grazing utilisation.

Table IE4. Management of new reseeds.

	Do's	Do not's
First 8 weeks	Graze at 2-3 leaf stage Spray weeds before grazing Nitrogen and P & K Slug pellets (if required)	Graze at high cover (> 1 400 kg DM/ha) Do not harvest for silage
Second grazing onwards	Graze at 1 200 – 1 600 kg DM/ha (6 - 8 cm) Re-spray weeds if necessary	Allow high covers to develop Graze in really dry or wet conditions
Autumn	Keep grazing at 1 200 – 1 600 kg DM/ha Graze off well before first winter (> 4 cm) Light slurry application	Overgraze or poach Apply excessive slurry
Second year	Ensure the new sward receives adequate nitrogen Monitor soil P and K status	Overgraze or poach

Graze the new reseed as soon as the plants do not pull out of the ground. Plants will normally be 6 to 8 cm high. It is especially important that autumn reseeds are grazed before the first winter.

The first grazing does not have to be completed by the main grazing herd; calves or young stock may be a better option, particularly during poor grazing conditions.

All the benefits of reseeding can be lost after sowing due to:
− Poor soil fertility – poor establishment and tillering
− Grazing at high grass covers or cutting for silage – tiller/plant death
− Weed infestation (especially docks) – loss of ground cover
− Pest attack (frit fly, leatherjackets and slugs) – tiller/plant death

Tillering

Tillering is the production of new grass plants by the main grass plant established from the seed. The process of grass tillering is critical for successful sward establishment. Tillering helps reduce the space available for weeds. To encourage tillering:
– Apply 40 kg N/ha 3-4 weeks after sowing
– Graze the reseed when it is about 6-8 cm high
– Continue to graze the reseed in the first year of production
– Avoid cutting the new reseed for silage in the first year (if possible)

Weed control:
– Weeds in new reseeds are best controlled when the grass is at the two- to three-leaf stage
– Docks and chickweed are the two most critical weeds to control in reseeds
– High populations of other weeds such as fat hen, charlock, redshank and mayweed can cause problems
– It is essential to control docks and chickweed at the seedling stage; this is achieved by applying a herbicide before the first grazing
– To achieve the best lifetime control of docks in a sward, eradicating the dock at seedling stage in a reseed is the best opportunity
– Herbicide choice for dock control will depend on the presence of clover in the reseed (see Herbicide Guide)
– Chickweed can be a problem particularly where regular grazing is not expected to take place (silage fields), therefore herbicide choice is important
– You should consult your local advisor or merchant representative for correct herbicide choice
– Remember to keep the prescribed cross-compliance records and follow the instructions on the product label

▸▸ Reseeding investment

Reseeding is one of the most cost-effective investments that can be made on a grassland farm. Projected costs are shown in Table IE5.

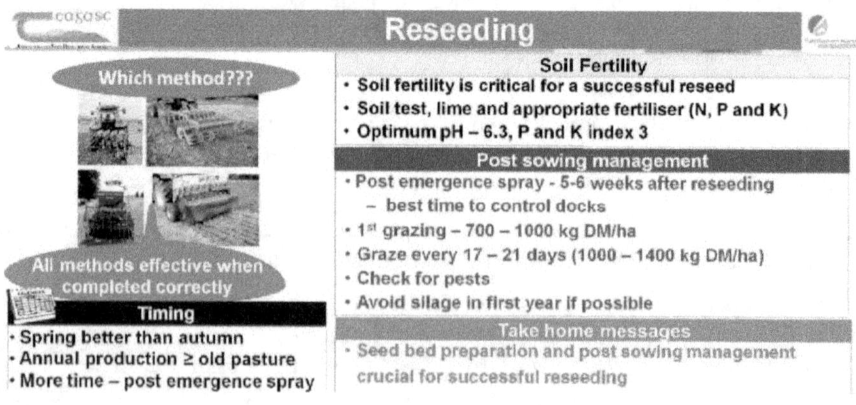

Figure IE2. Reseeding.

Table IE5. Projected costs.

	€/acre
Spraying	10
Glyphosate (Gallup 360) (Round-up (2 litres/acre)	16
Ploughing (€30)/ Till & sowing (one pass) (€30)	60
Fertiliser (2 bags × 10:10:20)	37
Fertiliser spreading	10
Levelling	10
Rolling	10
Grass seed	60
Post emergence herbicide sprays	30
Spraying	10
Costs (ex- post emergence sprays)	**253**

▶▶ PastureBase Ireland

Technologies that enable data-informed decision-making on the farm can help increase farmers' confidence and greatly improve grassland management. Huge leaps have been made in developing decision support tools to improve resource farm efficiency, profitability and sustainability. The primary objective of most of these tools is to increase the information available to assist in farm-management decision making as well as to collect and collate large amounts of data in a centralised database. Teagasc launched PastureBase Ireland (PBI) – an online grassland management decision support tool – in January 2013 and Grass10 will see the roll-out of the new PastureBase Ireland website as a key component of the campaign. Upon entering data from their own farm (e.g. grass measurements), the platform provides real-time and customised grassland management advice to farmers to assist their decision-making. These reports are developed in such a way that allows farmers to benchmark their individual farm with farms in their discussion group or in their region. The data accumulated to date indicate that PBI participating farms have achieved improvements in grass DM production and grazing management.

PastureBase Ireland is informing us that farmers need to have a good control of current grass supply in order to manage grass well. Grass cannot be managed correctly without knowledge of farm cover, grass demand and grass growth. The crucial point on any farm is utilising the feed resource produced on the farm.

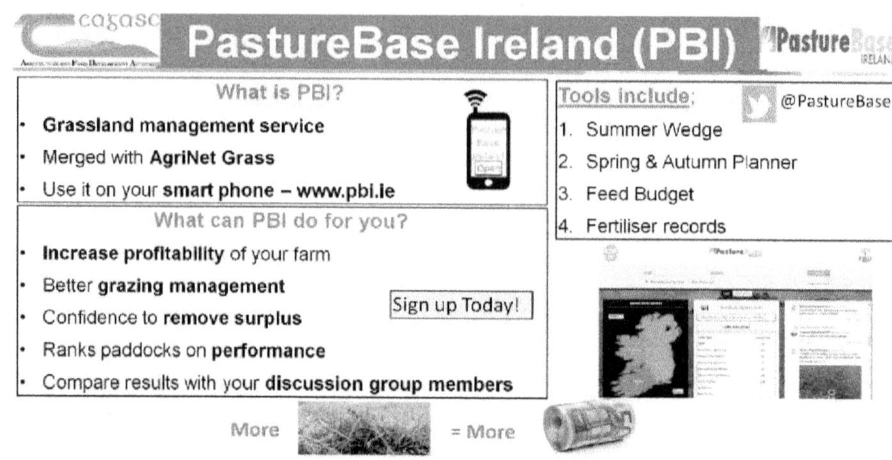

Figure IE3. PastureBase Ireland (PBI).

The Netherlands

Agnes van den Pol-van Dasselaar,
Leanne Bastiaansen-Aantjes, Hein de Kort

The description for the Netherlands is partly based on Van Dijk *et al.* (2015) and van den Pol-van Dasselaar *et al.* (2015 and 2018).

▸▸ Introduction

The Netherlands is a small and densely populated country. There are about 16.8 million inhabitants on about 4.15 million ha. This corresponds to 406 inhabitants km^{-2}. The area of cultivated land was 2.3 million ha in 1960, but in just over 50 years this area decreased by about 500,000 ha due to conversion into urban, industrial and nature areas.

Agriculture is of considerable economic importance for the Netherlands. The country is more than self-sufficient in many agricultural products. The total value of exports of all agricultural products amounted to about 75 billion euros annually in recent years. Cattle husbandry contributed significantly to exports through dairy products, cattle and beef.

The baseline for dairy production is good. The infrastructure is good (harbours, airports, roads), and therefore materials that are needed can easily be accessed. The Dutch agrosector is important for the Dutch economy; together with France, it ranks second on the list of agricultural product exports, behind the United States. The Netherlands is a world-leading exporter of milk and dairy products. Furthermore, there is a good knowledge infrastructure in the Netherlands (universities, schools, advisory services, farmer associations) that enables farmers to quickly assess up-to-date information to optimise their farm management.

▸▸ Dairy farming systems in the Netherlands: developments and characteristics

Today, dairy farms in the Netherlands are usually specialised, with their main activity being dairy production. In the Netherlands, forage production and grassland management have undergone substantial changes over the last 50 years. Yields, quality and utilisation of crops has increased due to improved grassland management, fertilisation and breeding. The average number of dairy cows per farm increased tenfold to about 85, average milk production per cow doubled to just over 8,000 kg milk/cow, average milk production per ha tripled to about 15,000 kg/ha and the number of dairy farms declined tenfold to about 18,000. This is illustrated in Table NL1, where a number of key parameters for the period 1960–2015 are shown.

Table NL1. Developments in dairy cattle husbandry in the Netherlands (Van Dijk *et al.*, 2015).

	1960	1975	1985	1995	2005	2010	2013
Agricultural area (×1 000 ha)	2317	2082	2019	1965	1938	1872	1848
Grassland area (×1 000 ha)	1327	1286	1083	1048	976	951	932
Forage maize area (×1 000 ha)	0.5	77	177	219	235	229	230
Number of dairy farms (×1 000)	183	91.5	58	37.5	23.5	19.3	18.5
Number of dairy cows (×1 000)	1628	2218	2367	1708	1433	1479	1553
Number of cows/farm	8	24	41	46	61	75	84
Kg milk/cow/year	4205	4650	5330	6610	7550	8000	7990
Kg concentrate/cow/year	800	1590	2280	2210	2020	2060	
Kg milk/ha/year (×1 000)	5.5	8.86	12.51	12.02	12.56	14.07	
Kg milk/farm/year (×1 000)	37	112.5	217	302	460	597	671
Kg milk in Holland/year (mill. tons)	6.7	10.3	12.5	11.3	10.8	11.9	12.2
Kg milk/h labour	8	37	72	89	128	150	
Dairy cows/ha grass and forage crops	1.2	1.6	1.8	1.3	1.2	1.2	1.3

Developments in milk production per cow and numbers of dairy cattle are shown in Figure NL1 at the national level for the Netherlands. Until spring 2015, the EU milk quota system limited the maximum amount of milk produced per country. The total number of dairy cows decreased following the introduction of the milk quota system in 1984 and the average milk production per cow increased.

The method of milking changed significantly, going from milking by hand to milking robots. Currently, around 25% of Dutch dairy farmers use milking robots (KOM, 2019). Amongst other effects, this led to a large increase in the number of dairy cows that can be managed by one person.

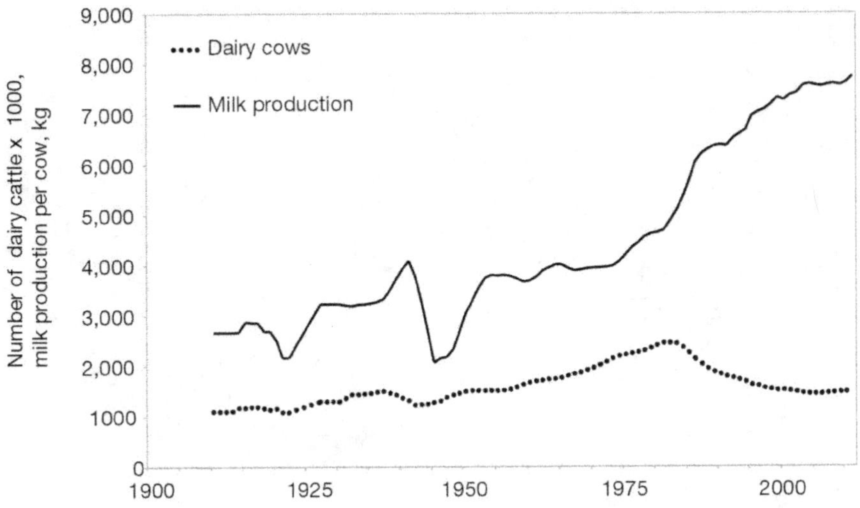

Figure NL1. Changes in the number of dairy cows and in milk production per dairy cow in the Netherlands from 1910 to 2011 (data from CBS; Van den Pol-van Dasselaar *et al.*, 2015).

▶▶ Soils

Even in relatively small land areas, such as the Netherlands, large regional differences in soil quality exist. Soil formation is strongly influenced by the North Sea and the Rhine and Meuse rivers that flow through these regions, as well as by climate and human intervention. Approximately 60% of the Netherlands is situated below sea level (-1 to -7 metres) and protected by dykes, dams and dunes. Clay soils are mainly found near the sea, rivers and in areas that were reclaimed from the sea. Peaty soils are found in the western and northern parts of the country. Sandy soils are mainly situated in higher parts of the country which are above sea level in the east and south.

Regional differences in soil type are reflected in regional differences in soil quality. Part of this fertility was inherited from the sea and the river deltas, part is man-made (e.g manure applications). The soils in the eastern and southern part of the Netherlands were originally mostly poor sandy soils. Current fertility status of soil organic matter content and soil P content in the Netherlands is shown in Figure NL2. The peaty (marine) soils in the west and the north can be recognised as areas with a relatively high soil organic matter content. Soil P content is generally high.

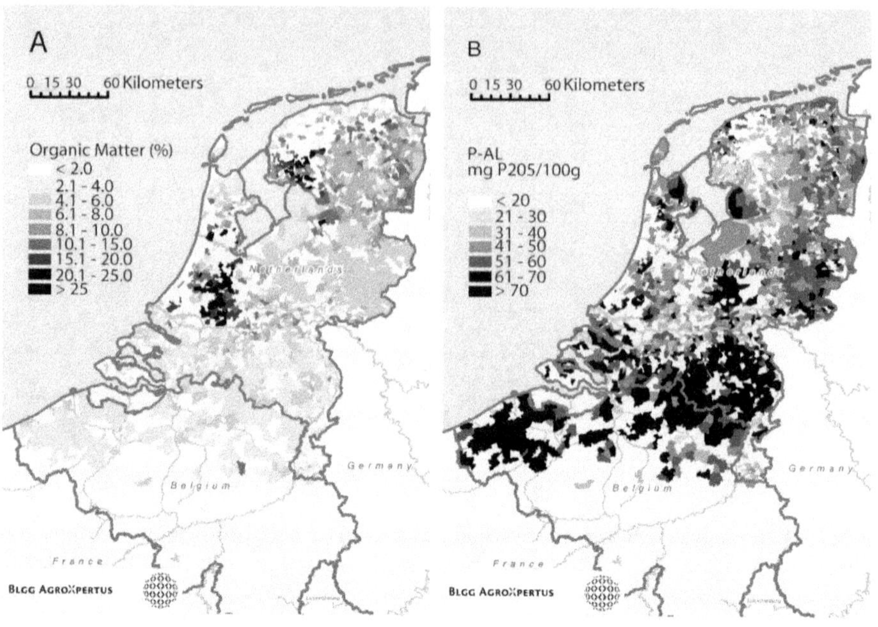

Figure NL2. Examples of soil quality in the Netherlands: a) soil organic matter content (%), and b) P content, P-Al, mg P_2O_5/100 g. Source: Blgg AgroXpertus, average of >70,000 samples in the Netherlands (Van den Pol-van Dasselaar *et al.*, 2015).

▶▶ Grassland areas

Grassland is the most important agricultural land use in the Netherlands (in the last years between 900,000 and 1,000,000 ha). In the last decades, the grassland area gradually declined to the present area, mainly due to the conversion of grasslands to grow maize. Furthermore, a lot of grassland areas were and still are being converted to extend roads and urban, industrial and nature conservation areas. The majority of grasslands is used for dairy production. Part of the grassland area is also used by sheep, beef cattle, horses etc.

Grasslands can be found on all soil types in the Netherlands: clay, sand, peat and loess. The peat areas, with relatively high groundwater levels (mainly in the western and northern part of the Netherlands), are predominantly used as grassland. Before 1970, the majority of the grassland was permanent grassland, i.e. more than five years old. Crops were hardly ever rotated. However, during the last 20 years, permanent grassland coverage has dropped to about 70%, mainly due to rotation with maize or exchange of land with arable farmers for cultivation of potatoes, flower bulbs etc.

Currently, an average of about 10% of all grassland is resown annually. The actual annual resown area is related to damage caused by drought or frost. In order to limit N leaching, it is obligatory to resow in spring on sandy soils. For grassland improvement, mainly mixtures of good and productive grass species and varieties are used. There are many mixtures available, which mainly contain perennial ryegrass varieties, both mid-late and late heading. Sometimes timothy and clover seeds are used as well. The use of specific mixtures is increasing, e.g. mixtures for mowing only, for grazing and mowing, with additional structure and with clover. There are also mixtures for sod sowing and/or temporary grassland. Diploid varieties are increasingly being replaced by tetraploid varieties.

▸▸ Grass yield and grass quality

The temperate maritime climate influenced by the North Sea and the Atlantic Ocean leads to cool summers and moderate winters. Daytime temperatures vary from 2°C to 6°C in winter and 17°C to 20°C in summer. The annual rainfall of 700 mm to 800 mm is evenly distributed over the year. These are good conditions for abundant grass growth in the growing season (April to October). Gross yields of 16 to 18 tonnes dry matter (DM) per year are common. However, the gross yield (i.e. the grass yield that is grown on the field) is less important than the net yield (i.e. the herbage that is either taken up by the dairy cow or mechanically transported from the field). For determination of net yield, grazing losses and harvesting losses should be deducted from the gross yield. Aarts *et al.* (2005) estimated the net yield of grasslands in the Netherlands and calculated an average of 10.4 tonnes DM per year (9.6 for peaty soils, 10.3 for clay soils, 10.4 for wet sandy soils and 11.5 for dry sandy soils). Variation among farms is considerable.

Table NL2. Median values and mean annual change (indicated by slope *b*) of grass quality; grass samples taken from grass silage in the Netherlands. The regression coefficient indicates the mean change of herbage quality per year for the period 1996 – 2009 (Reijneveld *et al.*, 2014).

Herbage characteristics[a]	Median	Slope *b*	R^2
Dry Matter	435	n.s.	n.s.
Crude Protein	165	-2.43**	0.52
Crude Fibre	242	n.s.	n.s.
Crude Ash	101	-2.53**	0.69
S	3.0	n.s.	n.s.
P	4.0	n.s.	n.s.
K	35	-0.30*	0.32
Mg	2.4	n.s.	n.s.
Ca	4.7	n.s.	n.s.
Na	2.5	n.s.	n.s.
Se	52	8.33**	0.51

[a] DM in g/kg, Se in mg/kg DM, all other in g/kg DM; * $P<0.05$, ** $P<0.01$, n.s. = not significant

Trends in average grass quality during a period of 15 years in the Netherlands are shown in Table NL2. Crude protein content and crude ash concentrations of grass decreased during the last years; those of K decreased and Se increased. The decrease coincided with a decreased fertiliser application. The increase in Se content can be explained by the increased use of Se-containing fertilisers (Reijneveld *et al.*, 2014).

▶▶ Rations

Dairy cattle rations in the Netherlands are in general characterised by relatively large amounts of supplementation, mainly maize silage, grass silage and concentrates. Silage maize is mainly grown on sandy and clay soils. Farms in those areas usually supplement more silage maize than farms in areas where not a lot of maize is produced. During the last decades, the total DM intake from grass has remained relatively stable, but the DM intake from grass silage has increased at the expense of grazed grass. This is partly explained by the fact that the herd size of the dairy farms has increased while the area around the farm that could be grazed has not increased to the same extent.

▶▶ Fertilisation

The government set rules for storage and application of manure in order to limit NH_3 emission and N leaching to ground and surface water. Amongst others, this means covering of manure storage, changes in method of manure application (sod application or manure injection to reduce emissions) and the period of manure application (during the growing season only) and limits the amount of animal manure per ha. Differentiations were made for various soil types and land use with respect to maximum fertilisation. In addition, standards for use and losses of N and P per ha grassland and ha forage maize were set.

The Netherlands is allowed by the EU derogation arrangement to apply 250 kg N from animal manure per ha grass and forage maize. The total fertiliser standards for the amount of N as well as P have been gradually become stricter in recent years, in particular for those areas with sandy soils where the nitrate content in the ground water still is too high. As a consequence of the maximum allowable amount of animal manure, application of N and P fertiliser sharply decreased in practice. The amount of applied fertiliser P/ha also decreased (Figure NL3). From 2014 onwards due to derogation regulations, application of fertiliser P on grass and forage maize is forbidden.

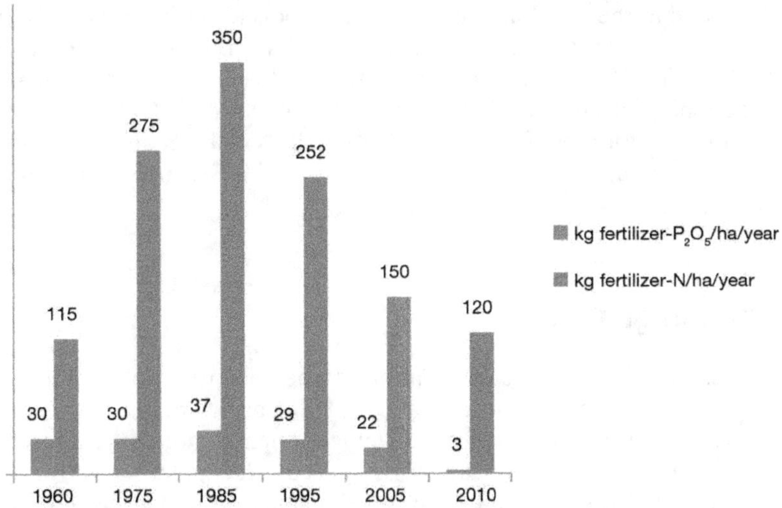

Figure NL3. Fertiliser P and N use/ha grassland/year in the Netherlands during the period 1960 – 2010 (Van Dijk *et al.*, 2015 and van den Ham A., LEI).

▸▸ Nutrient losses

High output dairy farming systems in the Netherlands are generally characterised by high fluxes of nitrogen (N) and phosphorus (P) through the systems. These elements cycle through the system by transfer between the components of the farm, i.e. from crops/feed to the herd, from the herd to manure, from manure to soil and from soil to crops/feed. Inadequate nutrient management of these intensive nutrient flows may cause high losses to the environment, which puts the quality of water, air and nature under pressure. Moreover, it reduces resource-use efficiency because not only exports as milk and meat but also losses from the systems are replenished by purchased feeds and fertilisers. From the mid-1980s onwards, efficient use of fertilisers with minimal losses to the environment was promoted and research efforts were focused in that direction.

The De Marke experimental farm was set up with the aim of exploring and demonstrating the possibilities to produce milk at an intensity of 12,000 kg milk/ha without violating strict environmental standards. Optimised mineral management on the pre-designed farming system resulted in a strong reduction of N and P surpluses compared to common practice. In the various management systems that were explored since 1993, N surpluses at farm level amounted to 98-165 kg/ha. P surpluses ranged from 0-6 kg/ha (Verloop, 2013). In the 'Cows & Opportunities' project, research on improving nutrient management was extended to commercial dairy farms on various soil types (Oenema, 2013).

The research has led to a strong decrease in mineral losses to the environment in practice in the years since. Moreover, it led to more insight into the flows of minerals at farm level and to the development of practical tools for farmers. For example, the Annual Nutrient Cycle Assessment (ANCA) model was developed in the Netherlands (Aarts *et al.*, 2015) to give farmers insight into the impact of their management on the functioning of nutrient cycles. From 2015 onwards, ANCA has served as a licence-to-produce for dairy farms in the Netherlands with a manure surplus (which is about 70% of the number of farms). It ensures that losses are minimised as much as possible.

▸▸ Trends in grazing

Grazing of dairy cows has several advantages, such as more possibilities for dairy cows to express their natural behaviour and the contribution to the image of the dairy sector. However, it also has certain disadvantages, such as more nitrate leaching and a less balanced diet (Van den Pol-van Dasselaar *et al.*, 2008; Hennessy *et al.*, 2015). It even affects the quality of the milk, since grazing increases the levels of unsaturated fatty acids in milk and meat (e.g. Elgersma *et al.*, 2006). The percentage of dairy cattle being grazed decreased from 90% in 2001 to 65% in 2015, but recently increased again (Figure NL4).

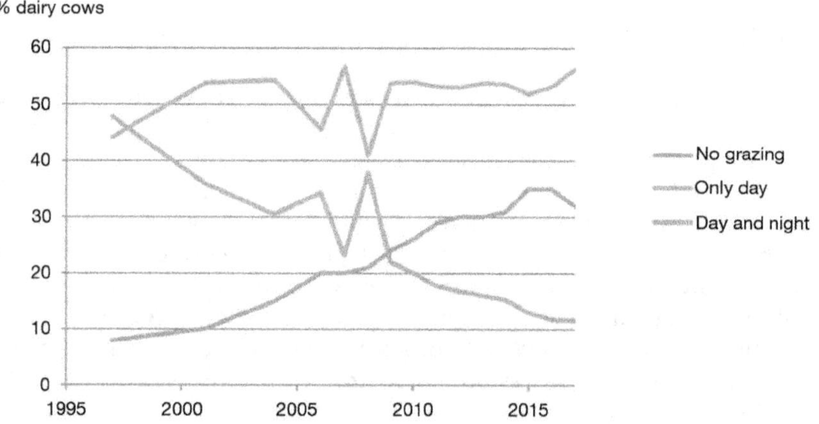

Figure NL4. Percentage of dairy cows grazed in the Netherlands during the period 1997 – 2017 (CBS, 2018).

Grazing is not evenly distributed over the Netherlands (Figure NL5). In general, there is less grazing in the south and the east of the country.

The decrease in grazing between 2000 and 2015 in the Netherlands can be explained by the following:
– To control rations and optimise grassland utilisation (when fed on grass only, DM intake is sufficient to meet the requirements of maintenance and milk production of 22-28 kg milk per day per cow)

– Grazing platform that does not increase, while herd size increases (see Figure NL6; Figure NL6 also shows an increasing percentage of grazing for larger herds in the last few years)
– Increased use of automated milking systems
– Grazing "can't be sold" (when cows are fully housed, more machinery and concentrates are needed)
– Need to reduce mineral losses
– Labour efficiency

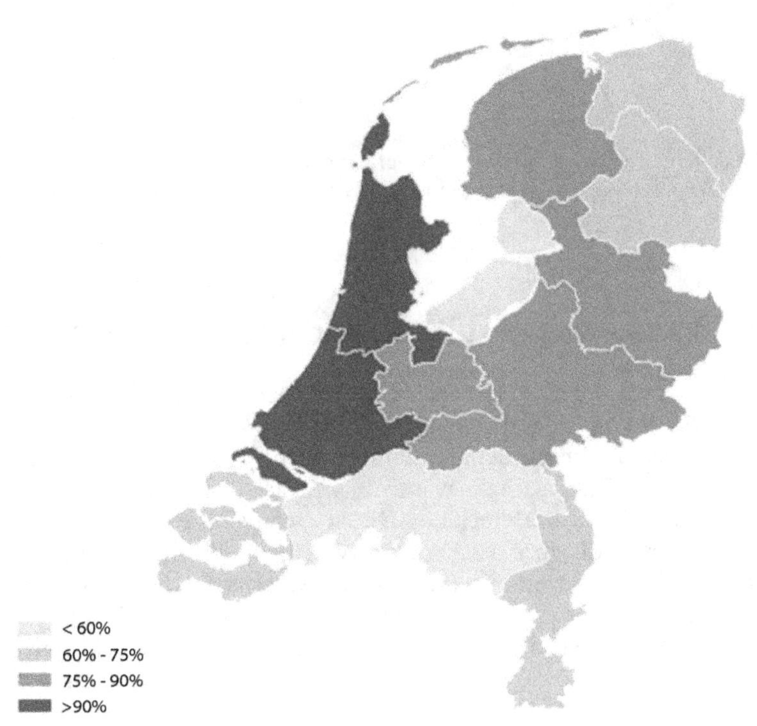

Figure NL5. Grazing in different parts of the Netherlands (CBS, 2018).

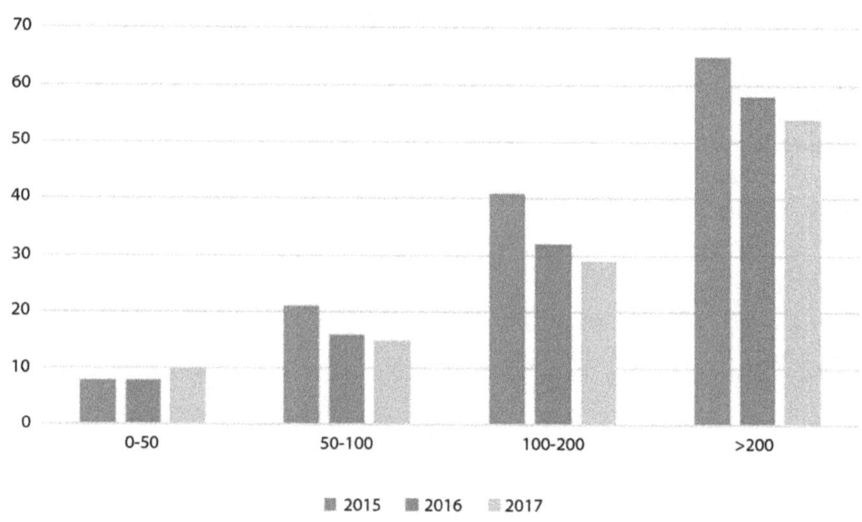

Figure NL6. Effect of herd size on zero-grazing in 2015, 2016 and 2017. Larger herds increased their percentage of grazing in recent years (CBS, 2018).

▶▶ Grazing as a societal issue and grazing premiums

Over the last decade the decrease in grazing has become a societal issue in the Netherlands. In Dutch politics and society, there is a broad interest in promoting cows with access to pasture. Public debates emphasise the high perceived value of grazing for animal welfare. The grazing cow is even seen as an icon of the Netherlands and part of the country's cultural heritage. Dutch society has expressed concern about less grazing. As a result, in 2012 a voluntary agreement, the 'Grazing Treaty', was signed (Figure NL7) by several partners in the Netherlands with the aim of stabilising the percentage of dairy farms that practise grazing. To date, around 83 parties have signed the agreement, which reflects the importance of grazing in the Netherlands. Signatory parties include representatives of dairy farmers' associations, the dairy industry, the feed industry, the milk robot industry, banks, accountants, the semen industry, veterinarians, cheese sellers, retailers, NGOs, nature conservation groups, government, education and science. As part of the treaty, many stimulating initiatives took place. The most prominent was the introduction of a grazing premium of one to two cents per kg that is provided by the dairy industry to farmers that practise grazing for at least 120 days per year for at least six hours per day. 'Pasture milk' is processed in separate milk streams and the majority of the Dutch supermarkets only sell such milk.

Grazing became an issue even in parliament in 2017 when a number of political parties suggested the requirement to make grazing obligatory. Other parties were confident that the 'Grazing Treaty' would prevent a further decrease in grazing and, as such, the treaty prevented the obligation. At the end of 2018, it was shown that the percentage of dairy farms with grazing is increasing, which is seen as a success for the treaty (Duurzame Zuivelketen, 2018) (Figure NL8).

Figure NL7. Signing of the 'Treaty Grazing'. Source: Duurzame Zuivelketen.

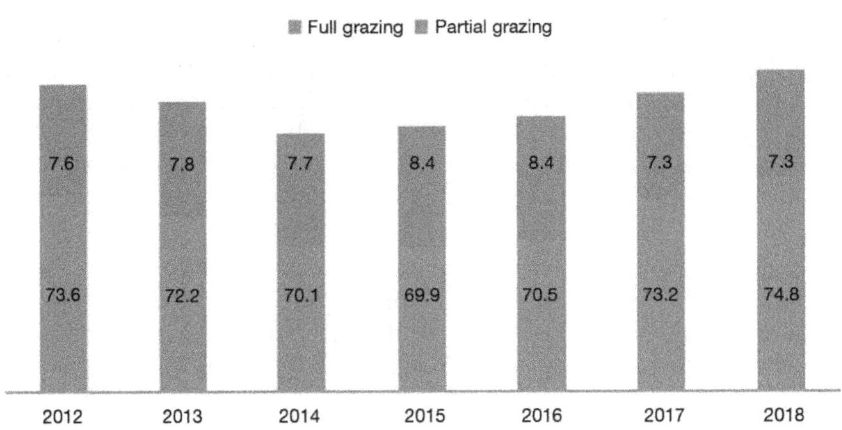

Figure NL8. Percentage of Dutch dairy farms that practise full grazing (in blue; at least 120 days of grazing and 720 hours of grazing per year) and partial grazing (in red) (Duurzame Zuivelketen, 2018)

▸▸ Grazing systems

For cattle, various grazing systems can be applied that vary in:
− Number of hours of grazing per day
  • Day and night grazing
  • Grazing only during the day, housing at night, feeding additional maize or grass silage and sometimes fresh grass (majority of Dutch farms)
− Number of days of grazing per paddock
  • Rotational grazing: cows are regularly moved to a new plot for grazing, varying from once or twice a day to each two to six days
  • Continuous grazing: cows graze for a longer period of time (three to six weeks to months) on a large paddock. The grass allowance of the paddock can be kept relatively stable by adapting the land area or by supplemental feeding.

Dutch dairy farmers mainly implement rotational grazing. However, continuous stocking is increasing. Recent trends are:
− Shortening the period of grazing on one plot for rotational grazing to one to two days per plot
− Introduction of compartmented continuous grazing on many farms

Compartmented continuous grazing (*Nieuw Nederlands Weiden* in Dutch) is an adapted set-stocking system for stocking rates up to 10 animals per ha in which the cows rotate on a daily basis between six compartments on one platform. Each day, cows are moved to a new compartment and in a period of five to six days, they rotate over five or six compartments. The (variable) sixth compartment is cut for silage to increase sward utilisation. This means that cows come back in the same compartment after five to six days. The average grass height in the compartments is kept constant (8 to 12 cm) so that daily regrowth is available for intake. The gap between daily regrowth and animal demand is filled with supplementation.

The grazing system was developed a few years ago in the Netherlands as a system that combines high grazing efficiency with ease of labour in dairy systems that have a high stocking rate and a high milk production per cow. The system is easy to implement, can be done on every farm, does not require a lot of labour or management skills and gives good results with respect to milk production and grass utilisation. The area on the farm that is available for grazing will optimally be used with this grazing system.

Next to grazing, farmers can keep their cows inside for the whole summer and feed them freshly cut grass (zero-grazing) or silage (summer feeding). Various combinations of systems can be found in practice as well. Each of the systems has pros and cons. The best system for a certain farm mainly depends on the farm's infrastructure, available manpower, number of cows, stocking rate and allocation of grassland. The best system can also change during the year. Consistent management is important for all grazing systems.

▸▸ Ensiling grass

Over the course of the years, various silage making systems have been used. In the last 10 to 15 years, 85% to 90% of all grass cut for winter forage has been ensiled, mainly as wilted silage. After a field period of two to three days, the grass is ensiled at a dry matter content of 35% to 45%. Wilting leads to a higher osmotic pressure in the grass cells which inhibits unwanted bacteria from developing in the silage. Wilting appeared to be the best and cheapest conservation method for young, protein-rich grass. Basically, for this ensiling method, additives are not needed and there are no problems with environmental pollution caused by silage effluent, while intake of the silage by cattle is quite good. However, quick ensiling is important (preferably in one day) as is airtight silage storage.

Silage additives are used to a limited extent. In the past mostly acids, salts or molasses were applied. In recent years mixtures of bacteria are mainly used. Additives are applied only if suboptimal conservation results are expected, for instance when the grass is not sufficiently dry, is high in protein or low in sugars, or when the field period lasted too long. High dry matter silages can heat up when they are opened. This problem can be prevented with sufficient compaction during ensiling, correct airtight storage and sufficiently rapid feeding of the silage. There are also special mixtures of bacteria that can restrict heating of the silage. Such mixtures are used in practice on a limited scale. Overall, 5% to 10% of wilted silages are treated with bacteria. Chopping of grass has a positive effect on the preservation and density of the silage due to the bruising and mixing during chopping.

Ensiling large bales (round and rectangular, with and without plastic covering) has also become popular in the Netherlands. It is estimated that 15% to 25% of grass is ensiled in this way. The method is particularly attractive for storing special lots separately and in case silage needs to be sold. In addition, it is not necessary to immediately transport the bales to the storage yard after pressing and wrapping them.

Many farms still use their own machinery for harvesting activities. However, increasingly, contract workers are involved who take care of loading and transport. Contract workers have large forage wagons, choppers and balers available and can execute all the activities at reasonable costs per ha. In general, contracting costs are lower than the farmer's own mechanisation costs.

Today, most silage is stored in large clamps on concrete surfaces or in bunker silos. In particular, the number of large bunker silos has increased in recent years. Advantages of those silos are the limited investments, correct storage and various machinery available for filling the silo and removing the silage.

▸▸ Outlook on the future of dairy systems in the Netherlands

An integrative approach for grassland management that is cost effective, environmentally sound and manageable is essential in the context of the development of large-scale dairy enterprises with highly productive healthy animals. New functions of high output dairy farming systems arise with corresponding revenue models, such as energy production, emissions trading, and the provision of other ecosystem services (e.g. cultural services). A variety of dairy farming systems may develop. Van den Pol-van Dasselaar *et al.* (2014) showed in a survey with approximately 2000 respondents from different European countries (mainly from Ireland, France, Belgium, the Netherlands, Poland and Italy) that the individual functions of grasslands, such as grazing, biodiversity, carbon sequestration, low-cost feed, landscape etc., are highly recognised and appreciated by all relevant stakeholder groups in Europe. All stakeholders considered that the large European grassland area is a valuable resource which is essential for the economy, environment and people. This could be exploited.

Further development of the dairy sector will require continuous development of people (education, training), tools (e.g. decision support systems) and techniques (innovations like sensors at cow and field level or techniques for manure refining). These developments are taking place in practice. Grasslands will remain an essential part of dairy farming systems, producing feed for dairy cattle. Grass production and utilisation should be stimulated by good grassland management, managing constraining factors like water shortage and using highly productive grass varieties and legumes. Soil (fertility) data, together with fertilisation registration, grass growth, weather data etc., collected at different resolutions, scales and time, and together with historical data, they could all be integrated in decision support systems. Multiple layers of information need to be analysed and assessed. This data assessment needs to increase grassland yield, improve herbage quality and ensure prudent use of nutrients. It is also essential to increase the net yields of grazed pastures by reducing grazing losses (trampling, urine and faeces) and by developing novel grazing systems for future dairy farms (large-scale, highly productive, highly automated) that are technically and socially feasible, economically viable and environmentally sound. The potential to improve grass yield is enormous as can be seen by the current variation in grass yield in practice. Optimal grassland use will lead to profitable farming with minimised environmental impacts while addressing demands from society such as animal welfare and grazing.

▸▸ References

Aarts H.F.M., Daatselaar C.H.G., Holshof G., 2005. *Bemesting en opbrengst van productiegrasland in Nederland*. Rapport 102. Wageningen, Plant Research International, 33 pp.

Aarts H.F.M., De Haan M.H.A., Schröder J.J., Holster H.C., De Boer J.A., Reijs J.W., Oenema J., Hilhorst G.J., Sebek L.B., Verhoeven F.P.M., Meerkerk B., 2015. Quantifying the environmental performance of individual dairy farms - the annual nutrient cycling assessment. *Grassland Science in Europe*, 20, 377-379.

CBS (2018) http://statline.cbs.nl

Duurzame Zuivelketen, 2018. https://www.duurzamezuivelketen.nl/nieuwsberichten/record-aantal-boeren-laat-de-koe-buiten-lopen/

Elgersma A., Dijkstra J., Tamminga S., 2006. *Fresh Herbage for Dairy Cattle – the key to a sustainable food chain*. Springer (194 pp).

Hennessy D., Delaby L., Van den Pol-van Dasselaar A., Shalloo L., 2015. Possibilities and constraints for grazing in high output dairy systems. *Grassland Science in Europe*, 20, 151-162.

KOM, 2019. http://www.stichtingkom.nl/index.php/stichting_kom/category/statistiek

Oenema J., 2013. *Transitions in nutrient management on commercial pilot farms in the Netherlands*. PhD thesis, Wageningen University, Wageningen, the Netherlands.

Reijneveld J.A., Abbink G.W., Termorshuizen A.J., Oenema O., 2014. Relationships between soil fertility, herbage quality, and manure composition on grassland-based dairy farms. *European Journal of Agronomy*, 56, 9-18.

Van den Pol-van Dasselaar A., Vellinga T.V., Johansen A., Kennedy E., 2008. To graze or not to graze, that's the question. *Grassland Science in Europe*, 13, 706-716.

Van den Pol-van Dasselaar A., Goliński P., Hennessy D., Huyghe C., Parente G., Peyraud J.-L., 2014. Évaluation des fonctions des prairies par les acteurs européens. *Fourrages*, 218, 141-146.

Van den Pol-van Dasselaar A., Aarts H.F.M., De Caestecker E., De Vliegher A., Elgersma A., Reheul D., Reijneveld J.A., Vaes R., Verloop J., 2015. Grassland and forages in high output dairy farming systems in Flanders and the Netherlands. *Grassland Science in Europe*, 20, 3-11.

Van den Pol-van Dasselaar A., Becker T., Botana Fernández A., Peratoner G., 2018. Social and economic impacts of grass based ruminant production systems. *Grassland Science in Europe*, 23, 697-708.

Van Dijk H., Schukking S., Van der Berg R., 2015. Fifty years of forage supply on dairy farms in the Netherlands. *Grassland Science in Europe*, 20, 12-20.

Verloop J., 2013. *Limits of effective nutrient management in dairy farming: analyses of experimental dairy farm De Marke*. PhD thesis, Wageningen University, Wageningen, the Netherlands.

Belgium

Alain Peeters, Benoît Delaite, Daniel Jacquet and Patrik Gauder

This section is largely based on the following publication:

Peeters A., 2010. Country Pasture Forage Resource Profile: Belgium. Food and Agriculture Organisation of the United Nations (FAO), 75 pp.

However, this text has been fundamentally revised, completed and updated.

▸▸ Introduction

Land area and land structure

In Belgium, agriculture occupies almost 60% of the land mass, woodland areas about 20%, and urban areas, including residential, industrial and green areas as well as communication ways, about 20% (Table BE1). There has been a steady decrease in the agricultural area (AA) to the benefit of urban areas since the 19th century. The permanent grassland area increased substantially in the second half of the 19th century, but decreased slightly after the 1970s and is still decreasing.

Table BE1. Land use in Belgium and evolution between 1834 and 2017 (FPS Economy, 2008; FPS Economy – DGSEI, 2009, Statbel, 2018).

	1834	1990	2000	2007	2017
Land area (km^2)	29,456	30,278	30,278	30,278	30,278
Agricultural area (km^2)	19,172	18,302	17,653	13,703	13,292
Arable land (km^2)	15,087	7,594	8,631	8,396	8,357
Permanent grassland (km^2)	3,468	5,786	5,069	5,073	4,678

Socio-economic and structural aspects of agriculture

There were about 35,900 farms in Belgium in 2017. Two thirds were located in the Flemish Region, one third in the Walloon region. Farms are more than twice as large in Wallonia (57 ha) compared to Flanders (26 ha). In Wallonia, the industrial revolution began in the early 19th century and created jobs outside the agricultural sector, leading to a transfer of labour from agriculture to industry (coal mines and steel factories at that time). This permitted a faster farm size increase compared with Flanders, where the industrial revolution did not begin until the mid-20th century. Flemish farmers had to intensify their production if they wanted to survive on their small farms in this densely populated region. In the second half of the 20th century, they specialised in flower, vegetable, pig, poultry or dairy productions. These productions, especially the four first, can provide significant incomes on very small surface areas. In Wallonia, the lack of space was less critical; farms specialised in cereals, sugar beet, beef and dairy productions. The two development models are thus radically different. Most Belgian farms (about three quarters) have permanent grasslands, and half grow forage crops and/or raise cattle.

Flemish plain. Left: Grasslands, avenue of poplars (*Populus*) and some pollarded willows (*Salix alba*). Right: Old pollarded willows (*Salix alba*). Source: ILVO – Merelbeke.

Ardennes. Left: Plateau with a village, grassland plots, arable land (in the background) and farm buildings (in the foreground). Middle: Temporary grassland (grass/clover mixture) managed by cutting for making haylage or silage. Right: Belgian Blue cows grazing. Spruce (*Picea excelsa*) forest in the background. Source: S. Cremer.

The sector accounts for a bit more than 0.5% of the Belgian GDP in 2017 (Statbel, 2018) but it has a more significant importance when the upstream (e.g. machinery, fertiliser and pesticide productions) and downstream (e.g. mill and sugar industry, slaughterhouses, animal feeding and dairy industries) sectors are considered. The

190

economic importance of the entire chain is often estimated at about 10% of the country's GDP (FPS Economy, 2007). With regard to exports, the whole agricultural sector (animal and plant products, food products, beverages and tobacco) represented 11.8% of the Belgian GDP in 2017 (Statbel, 2018).

▸▸ Ruminant livestock production systems

Recent history and context

Dairy production increased from an average annual production of 4,500 l/cow in the 1970s to 7,800 l/cow in about 35 years. Certain herds and cows are currently reaching annual production of 10,000 to 12,000 l/cow. This regular increase of about 1.3% per year in dairy performances was made possible thanks to an international breeding effort of the Holstein-Friesian breed. This led to a decrease in the populations of many dual-purpose breeds (e.g. Belgian Red, Red and White, dual-purpose Belgian Blue). Yield increases also led to changes in animal feeding. More concentrates were used at the expense of the proportion of green forage. The implementation of milk quotas slowed this trend because the control of the volume of production required a decrease in production costs. This was achieved by better utilisation of green forage, which is less expensive than concentrates. Values of 3,000 to 4,500 litres of milk per cow and per year are regularly obtained on the basis of green forage (forage maize and grassland). The decrease in production costs was also achieved through higher per-cow production. The milk quota of each farm has been fulfilled with a decreasing number of cows, which in turn decreased the proportion of maintenance feeding needs and increased the proportion of production needs in total feeding costs. The CAP reforms of 1992 and 2000 resulted in a significant decrease in cereal prices (about 50%), which once again encouraged dairy farmers to use cereals in animal feeding. Moreover, farmers tended to use more maize silage at the expense of grass grazing and grass silage when dairy cow production exceeded a certain threshold (roughly above 6,000 l/cow). They did not trust grass quality and grass intake potential of their high-yielding cows, especially because of rainy weather and unfavourable temperatures. They generally kept cows indoors part of the time or systematically complemented grassland grazing with maize silage. This led to a decrease in the grassland proportion in the agricultural area. This trend was very strong in Flanders (polders, sandy region, sandy-loamy region, Campine/De Kempen). In the grassland regions in south-eastern Belgium, farmers' strategies differed slightly; they used less maize but increased concentrate use. They also attempted to reduce production costs through better use of grazing and grass silage.

In beef production systems, the Belgian Blue emerged as the almost only profitable breed in the 1970s. Grazing remained the basis of suckling cow systems and animals were fed in winter mainly with hay and haylage. Concentrates and maize silage were restricted mainly to bull fattening. Ox fattening nearly disappeared in favour of young bulls.

All these system changes had impacts on landscape and wildlife by reducing diversity and complexity. Farmers faced criticism for their negative effect on ground- and surface water quality. Measures had to be taken to decrease nitrate and phosphate pollution. Farmers had also to respect the Habitats and Bird directives, but these covered only a small part of the agricultural area in Belgium. The Nitrates Directive was also mandatory for farmers and had a significant influence on intensive livestock system farm structures and practices by regulating the stocking rate and organic nitrogen management. The agro-environmental measures gave farmers an opportunity to adopt concrete measures that can improve their impact on the environment.

In summary, the successive EU CAP reforms led to 'modernisation' of the sector, farm size increases, a dramatic decline in the number of farmers, production specialisation, intensification of grassland and animal husbandry, an increase in production volume, increase of grassland and animal yields, a reduction in legume use, a reduction in grassland area and its proportion in the agricultural area, and finally a reduction in the diversity of landscape, grassland species and communities, and domestic animal breeds. The implementation of milk quotas reduced the number of dairy cows, which led to a stocking rate decrease in some cases or the development of beef production systems in complement to dairy systems in most cases.

The animal sector as a whole

Belgian agriculture is strongly oriented towards meat and dairy production. Between 1997 and 2007, there was a decline in beef meat production and an increase in pork meat production. Milk production was fairly constant due to EU milk quotas restricting production. However, after the milk quotas were phased out, Belgium's total milk production rose again between 2007 and 2017 from about 3,000 to about 4,000 million litres (Statbel, 2018).

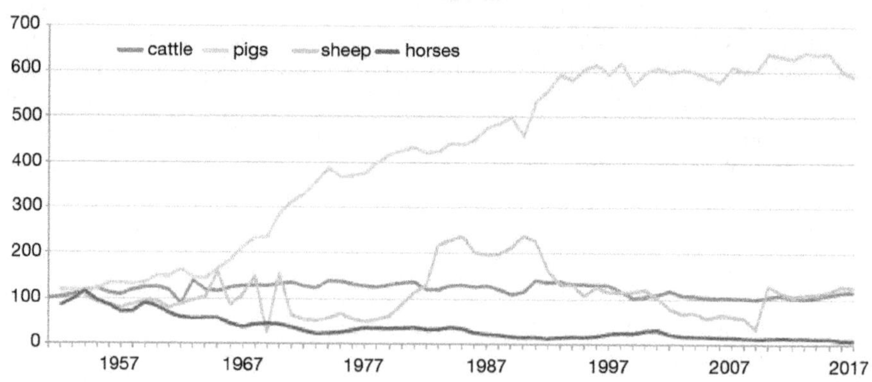

Figure BE1. Evolution of the number of slaughtered animals (%) (1951-2017) (1951 = 100) (Statbel, 2018).

Since 1990, the number of cattle and sheep decreased while pigs and poultry numbers increased. Figure BE1 shows the spectacular increase of pig populations, the stability of cattle and sheep populations and the decline of horses between 1951 and 2017. All species numbers increased per farm. Between 1990 and 2004, these animal numbers per farm evolved from 56 to 86 heads for cattle, from 332 to 785 heads for pigs and from 1,851 to 6,575 heads for poultry (Genot, 2005).

Intensification levels of livestock production can be defined by stocking rate, i.e. the ratio between livestock units (LSU) and the surface area (ha) of grassland and forage crops. In Belgium, they are considered as 'medium' ($1.6 \leq$ stocking rate < 3) for cattle and sheep and 'high' or 'very high' (high: $3 \leq$ stocking rate < 7; very high: stocking rate ≥ 7) for pigs and poultry. Compared with the European average, even cattle and sheep intensification levels are high or very high.

Dairy farming is traditionally concentrated in lower and central Belgium as well as in the Liège grassland region and upper Ardennes. The sandy region, the sandy-loamy region, the loamy region, the Campine/De Kempen and the Liège grassland region include almost 80% of the national dairy cow population. Suckling cows are concentrated in central and upper Belgium (Namur and Luxemburg provinces). The sandy-loamy region, the loamy region, the Condroz, the Famenne, the Ardennes and the Jurassic regions include about 80% of the national suckling cow population. Dairy cows are thus more abundant in the north and west of the country, and suckling cows in the south and east. Most dairy cows are bred at low altitudes. In higher altitudes, suckling cows are dominant. The Atlantic climate of the Flemish plain and the Campine/De Kempen creates very favourable conditions for dairy production because it provides good grass-growing conditions, with relatively regular growth throughout a prolonged growing season.

Dairy breeds in the Flemish plain. Left: Red and White cows by a canal. Right: Holstein cows in an urbanised landscape with poplar (*Populus*) plantations.

Source: ILVO - Merelbeke.

▸▸ Dairy production

Dairy cow herd size is typically 50 on specialised farms (Statbel, 2018). The national average production per cow is somewhat illogical in Belgium. It conceals significant variability in dairy performances between dual-purpose Belgian Blue cows that produce about 3,000 l/cow annually, dual-purpose Red and White cows that yield about 4,500 l/cow annually and specialised dairy Holstein-Friesian cows that produce much more. In 2017, average annual milk production per specialised dairy cow was 7,800 l and average values for milk fat and protein contents were 4.09% and 3.45%.

A high-yielding dairy cow ingests about 1,000 to 1,500 kg concentrate per year. With regard to green forages, the proportion of green grass, grass silage and maize silage in the diet varies considerably according to farming systems and regions. Winter housing is either a free-stall with a slatted-floor and a milking parlour or a tie-stall where milking is carried out in cow individual stalls. The first type of building is slurry-based while the second is farmyard manure (FYM)-based. Straw free-stall systems are developing. Dairy farms are usually smaller than beef farms and have a higher per-ha income.

▸▸ Beef production

Beef cattle herd size vary considerably according to regions and farming systems. Beef farms have about 30 suckling cows on average (2017) (Statbel, 2018). Mating is mostly natural in beef production. During the grazing period, each suckling cow herd on a farm includes one or several breeding bulls including in the herd of the tallest heifers. The calving period is concentrated between December and March. Male and female calves suckle their mother, which produces about 1,000 to 2,000 litres per year. Winter housing is either a free-stall or a tie-stall. Both are bedded with straw and produce FYM. In tie-stalls, calves are kept in pens and have access to their mother for a limited time per day. There is an increasing trend to separate calves and cows after calving in Belgian Blue systems. Young bulls never graze in this system since they are kept indoors up to the end of the fattening period just before slaughter. In traditional systems, cows and calves start grazing together in April-May. Calves drink milk, graze and have access to a feeding system for receiving concentrates in grassland, while cows only graze.

▸▸ Sheep and other ruminant productions

Sheep are raised in small numbers (about 151,000 in total in 2007) mainly for meat production. Sheep breeding is often a secondary or a hobby activity. Sheep are generally kept indoors during winter time. Lambing is concentrated between February and April for the production of 'herbage lambs' or 'grey lambs'. In this case, ewes and lambs graze together from April to August or September. Lambs can receive concentrates in grassland for a short period before slaughter at the end of summer.

Dry ewes graze until the middle of November. In this system, lambing occurs indoors in December and January. Lambs can suckle ewes and they also receive concentrates. They are slaughtered at three months old, in April, and so they never graze. Ewes graze alone from April to November.

There were about 29,000 goats only in 2007. Among the curiosities in this part of Europe, American bison and red deer can be cited. Most bison are raised in Wallonia and the Ardennes, and most deer in Flanders.

▶▶ Organic farming and stockbreeding

In the Flemish Region, the area under organic farming in 2017 was 7,367 ha (Statbel, 2018). This area corresponds to only 1.2% of the total agricultural area in Flanders. The same year, there were 468 Flemish organic farmers. Several socio-economic factors underpin the low development of organic farming in Flanders. Firstly, organic farmers have experienced difficulties in marketing their products, and secondly, traditional farmer advisory circuits have not supported organic farming.

In Wallonia, the number of organic farms and their total agricultural area are experiencing stronger growth, especially since 1996 (Statbel, 2018). At the end of 2017, organic farming covered 76,072 ha, or 10.6% of the Walloon agricultural area. This proportion is higher than the EU-15 average, which was 7% of the agricultural area in 2017 (Eurostat 2018). The same year, there were 1,625 organic farms in Wallonia. It is about 3.5 times more than in Flanders. In 2017, 78% of Belgian organic farms and 91% of the Belgian organic agricultural area were located in Wallonia.

▶▶ Pasture resources

Forage production systems

There are two main forage production systems in Belgium. The Flemish system is based on regularly resown grasslands and on annual forage crops (temporary grasslands and maize). The system in Wallonia is mainly based on permanent grasslands. Interestingly, the highest proportions of grasslands in the agricultural area are mainly observed in areas with low density human populations (provinces of Luxemburg, Namur and Liège).

Grasslands and forage crops

Table BE2 summarises typical Belgian yields of forage maize and fodder beet, compared to intensive-cut temporary grasslands and grazed permanent grasslands.

Table BE2. Comparison of typical yields of the main forage crops in Belgium (Deprez *et al.*, 2007).

	Forage maize	Fodder beet	Cut temporary grassland	Grazed permanent grassland
t FM/ha	50	120	50	57
%DM	30	13-19	20-30	15-20
t DM/ha	13-18	14-21	12-16	8-12

FM = fresh matter; DM = dry matter.

Permanent grasslands

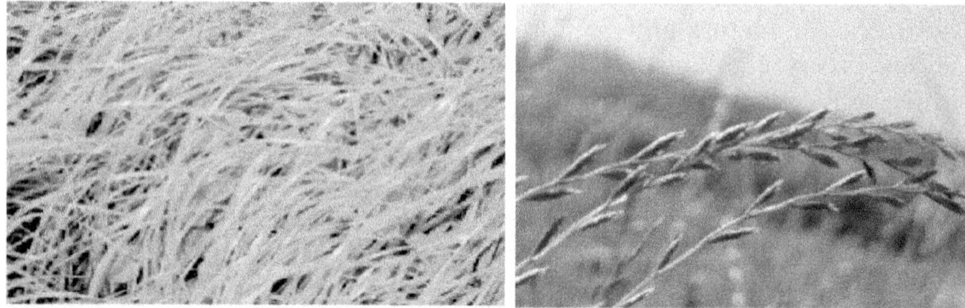

Grasses and legumes. Left: Perennial ryegrass (*Lolium perenne*) – white clover (*Trifolium repens*) mixture. Right: Perennial ryegrass (*Lolium perenne*) spikes. Source: ILVO – Merelbeke.

Legumes and grasses. Left: Red clover (*Trifolium pratense*). Right: Grass – Lucerne (*Medicago sativa*) mixture. Source: ILVO - Merelbeke.

Agriculturally-improved permanent grasslands include 15 to 20 species of higher plants on one ha (i.e. about 10 species on 100 square metres), among which grasses are dominant. The most common are: *Agrostis stolonifera* (on fresh and wet soils), *Alopecurus pratensis* (on fresh and wet soils), *Dactylis glomerata, Holcus lanatus, Lolium perenne, Poa pratensis* (especially on superficial or sandy soils that are dry during a part of the growing season) and *P. trivialis*. Some nitrophilous dicotyledons can be locally abundant: *Cirsium* spp., *Ranuculus repens*, *Rumex crispus*, *R. obtusifolius*, *Stellaria media*, *Taraxacum* spp. (especially in the Ardennes) and *Urtica* spp. *R. obtusifolius* is the main grassland weed. *Cirsium arvense* and *Urtica dioica* can also be problematic locally. The proportion of *Trifolium repens* in swards is usually low but it can be high when the defoliation frequency is high and usually when there is not too much nitrogen fertilisation. Other legume species are rare in intensively used permanent grasslands.

Species-rich cutting meadow in a nature reserve. Source: S. Rouxhet.

Temporary grasslands

Temporary grasslands can be mainly cut (one hay cut and aftermath grazing or two to three silage cuts plus grazing or three to four silage cuts), or mainly grazed (grazing only or grazing plus one to two silage cuts). Typical cut temporary grasslands are sown for one to five years. In recent years, they tend to be kept for a longer period, especially in grassland specialised regions in the south-eastern part of the country. Between 1960 and 1990 in the Flemish Region, mainly grazed temporary grasslands were established for about five years and then resown. Since 1990, there has also been a trend to convert them to permanent grasslands either with or without regular over-sowing.

Mixtures for mainly cut grasslands include one to five species. The most widespread species are *Lolium perenne* and *L. multiflorum* (about 90% of the seed market). Other sown species are grasses: *Dactylis glomerata*, *Festuca pratensis*, *F. arundinacea*, *Phleum pratense*, *Poa pratensis*, hybrids of *Lolium* spp. and sometimes *Lolium* x *Festuca*; and legumes: *Medicago sativa*, *Trifolium pratense*, *T. repens*. In the Ardennes, a traditional mixture includes *F. pratensis*, *L. perenne*, *P. pratense* and *T. pratense*. The frost resistance of *F. pratensis* and *P. pratense* is appreciated above 500 m asl. These species are added into this mixture as insurance against climate hazards. Simple mixtures of

L. perenne (usually pure or in mixture with *T. pratense* and/or *T. repens*), and simple mixtures of *L. multiflorum* (which may or may not be mixed with *T. pratense*) are the most frequent for the establishment of grasslands harvested for silage making. The mixture of *D. glomerata* and *Medicago sativa* was almost abandoned in the second half of the 20th century but its potential has received renewed attention from farmers from the lower and central areas of Belgium, as well as those from Condroz and the Jurassic Region.

Legumes in grassland swards

Trifolium pratense, *T. repens* and *Medicago sativa* are the three main forage legume species in Belgium. *T. pratense* and *M. sativa* are almost exclusively used in cut temporary grasslands. *T. repens* is mainly used for the sowing of long-lived grazed swards but its large-sized cultivars are also sometimes associated with medium-lived cutting mixtures. The capacity of N fixation and the high nutritive value and intake potential of these three legume species are increasingly appreciated by farmers after having fallen out of favour. Both characteristics can reduce production costs and thus increase farmer income. Surfaces devoted to the cropping of clovers and lucerne decreased dramatically from the 1960s in Belgium. They continued to decrease by 69% and 26%, respectively, between 1990 and 2000 (DGSEI, 1990 and 2000). Belgian farmers acknowledge the theoretical interest of legumes, but in their intensive grassland production systems, they tend to prefer the use of nitrogen fertilisation providing strong, consistent yields, even if it leads to sacrificing legumes. This practice has been maintained by the relatively low prices of nitrogen fertilisers and by some drawbacks of legume species. Despite undeniable breeding progress, persistence remains a problem, especially for *T. pratense*. A slow growth and a low nitrogen fixation in spring make grass/clovers, especially grass/*T. repens*, mixture less attractive than N fertilised pure grass swards. Despite a low mortality rate, bloat risk induced by *T. repens* in grazed swards is often overestimated by farmers.

▶▶ Grassland management and forage conservation

Grassland management systems are quite diverse in Belgium, although they are intensive almost everywhere.

Temporary or permanent grasslands

In the Flemish Region, grazed and mixed-used grasslands are traditionally resown, mostly on the same place, every four to five years. One fourth or one fifth of the grassland area of a farm is renovated at the end of summer by this technique. This system is still important in this region but there is a clear trend to increase the lifetime of the sward and to use more permanent grasslands.

In Wallonia, grassland production is mainly based on permanent swards. The most intensive dairy producers of the Herve country and the Upper Ardennes also prefer to keep permanent grasslands, but they try to improve their botanical composition by the introduction of *Lolium perenne* seeds.

Temporary grasslands that are used for the production of conserved forages are generally implemented into crop rotations. In lower and central Belgium, these short- and medium-lived swards are sown either with *Lolium perenne* for three to four years, either with *L. multiflorum* for one (ssp. *Westerwoldicum*) or two years (ssp. *multiflorum*). Mixtures are usually composed of several cultivars. Early cultivars of *L. perenne* are appreciated for silage making. They are mixed with intermediate cultivars, especially if the sward is sometimes grazed. Hybrid ryegrasses (*L. perenne* x *L. multiflorum*) are sometimes used for trying to combine the persistence and the excellent feeding quality of *L. perenne* with the high yielding potential of *L. multiflorum*. From the 1960s through the end of the century, the use of grass/legume mixtures declined considerably, but in the early 21st century, there was renewed interest in these types of mixtures, especially for *Medicago sativa* and mixtures with *Dactylis glomerata*. Some farmers understood that these forages are cheap sources of quality protein and that they can reduce fertiliser costs with biological nitrogen fixation.

Fertilisation

Nitrogen fertilisation is high in all regions even though it is limited by law. Across the entire Flemish Region, organic N fertilisation from animal manure is limited to 170 kg/ha on grasslands and most crops. Total mineral and organic nitrogen application must always be lower than 350 kg/ha per year on grasslands, but no N fertilisation is allowed on forage legumes (125 to 275 kg/ha on arable land). Across all of Wallonia, organic N fertilisation is limited on grasslands to 230 kg/ha per year (115 kg/ha on arable land). The total mineral and organic nitrogen application must always be lower than 350 kg/ha per year on grasslands (250 kg/ha on arable land).

The responsibility of farming in nitrate pollution of water tables has been frequently estimated at more than 80%. Farming in general and stockbreeding in particular are also responsible of emissions of NH_3 and N_2O in the atmosphere. The deposition of ammoniac in oligotrophic habitats like moorland, peatland and forests induces the eutrophication of these habitats and the disappearance of rare oligotrophic species. The entire functioning of these sometimes threatened habitats is disturbed because of these excessive N inputs from the atmosphere. On average, annual N deposition increased in Belgium from about 5 kg/ha in the 19th century to about 35 kg/ha at the end of the 20th century. In some areas of the Flemish Region, this annual N deposit reaches 50 to 80 kg/ha. N_2O is emitted mainly by wet and trampled N-fertilised grassland soils. This greenhouse gas contributes to climate change.

Phosphorus fertilisation was also excessive during the 30 to 40 years of the 'blind intensification period'. Many Flemish soils are saturated with phosphorus, and P leaching is even observed on sandy, easily leachable soils. On the loamy soils of Wallonia, P leaching does not usually occur, but many soils are so rich in P that the cessation of P fertilisation over 20 years of experiments did not cause yield reductions. Surface waters are polluted by soil phosphate runoff of mineral and organic P in both regions, which leads to eutrophication of rivers and aquatic biodiversity losses.

It has been calculated that P and K fertilisation is no longer necessary on dairy farm grasslands because inputs of these nutrients through concentrates are sufficient for compensating the outputs of meat and milk, if the nutrient cycle of these chemical elements through the spreading of slurry, urine and dung deposits during the grazing period is well managed. In specialised cut grasslands, exports of P and K from forage harvesting for conservation are important. They are increasingly compensated for through a concentration of organic manure application on this type of swards. In conventional farming, N fertilisers are still used on grasslands (except on well-managed grass/legume swards), but even the use of this nutrient has been strongly reduced through better information shared with farmers on the fertilisation value of organic manures. These manures were too often considered as wastes during the 1960–1990 period and their fertilisation value was not taken into account. This led to an overuse of chemical N fertilisation and pollution of the water tables and surface waters by nitrate. The EU Nitrates Directive played a key role in encouraging farmers to more precisely calculate the N balance on their farms and the N requirements of grassland plants. As a result, N and P fertilisations dropped significantly in the past 10 years.

A dense network of soil analysis laboratories provides fertilisation advice to farmers who use their services. In Wallonia, these laboratories are strongly subsidised by the public authorities.

Production

Belgium is one of Europe's areas of highest grassland production without irrigation (Peeters and Kopec, 1996). In cutting experiments (three to four cuts/year), annual yields of *Lolium perenne* are about 12-16 t DM/ha in lower and central Belgium (10-14 t DM/ha in the Ardennes). Annual yields of *L. multiflorum* are about 15-20 t DM/ha in lower and central Belgium (12-16 t DM/ha in the Ardennes). In a frequent defoliation regime (four-week intervals between cuts), average annual yields of about 9 t DM/ha were recorded in the Ardennes (Peeters and Kopec, 1996). Under farm conditions, annual yields of grazed swards typically range between 6-12 t DM/ha. In lower and central Belgium, annual yields of 10-12 t DM/ha are not rare in grazed swards.

The average stocking rate is about 2.4 LSU/ha of the forage area (temporary and permanent grasslands + forage maize) in Wallonia and about 3.2 LSU/ha in Flanders.

Grazing systems

Grazing systems have evolved significantly over time. They could be quite complex in the 1960s, but they are much simpler nowadays. Strip grazing was once frequently used in permanent grasslands, both for grazing forage crops established for a short period of several weeks between two main crops and for grazing longer-term temporary grasslands. Though effective, this system is now rare because it requires high labour input. Rotational grazing was the main system in the 1960s and 1970s. Mineral nitrogen fertilisation increased progressively since the 1960s and silage cuts were generalised in this system in the 1970s. Meanwhile, the intensive set stocking system was promoted. It was associated with higher nitrogen fertilisation levels compared to rotational grazing. In the 1960s, many farmers were afraid to spread mineral N fertilisers in the presence of cows in a paddock, but they quickly understood that this technique was safe. Most systems now fall somewhere between a pure rotational system with about 10 paddocks for the dairy cowherd and a pure set-stocking systems with a single paddock. These intermediate systems allow farmers to combine the advantages of both systems. They help minimise labour like in set stocking systems and optimise the flexibility of management that is characteristic of rotational grazing systems.

In dairy systems, the number of paddocks for dairy cows typically ranges from one to five in lower and central Belgium. In the specialised dairy regions of upper Belgium, Herve country and Upper Ardennes, rotational grazing is more strictly applied, and the number of paddocks is higher, from 10 to 15 on average. Three-day grazing periods per paddock and per grazing cycle are typical in this case. Two or three paddocks are usually devoted to heifers.

In beef systems, the organisation of grazing is more complex because of the existence of several herds of suckling cows and their calves grazing in separate groups of paddocks. It is frequent to have three to five suckling cow groups grazing separated grazing circuits. Each group, including the group of tall heifers, is accompanied by a bull. In the Ardennes, rotational grazing is more common. Each group grazes a small number of paddocks (four to six) for 10 to 20 days for each grazing period. Cows can be moved from one group to another for several reasons, such as when their calves are weaned.

In dairy systems, there is also a shift towards continuous (*ad libitum*) access of dairy cows to maize and/or grass silage during the grazing season. Since the dairy performance of cows continuously increases over time, farmers lose confidence in the potential of grasslands to feed high-yielding dairy cows properly. During unfavourable weather periods – rainy and cold or sunny and warm – farmers do not want to take the risk of an intake decrease that could led to a decline in dairy production and provide silage to these cows. Moreover, farmers tend to use more maize silage at the expense of grass grazing and grass silage when dairy cow production is above a certain threshold (roughly above 6,000 l/cow). Increasingly, access to silage is no longer restricted in many cases. This trend is reinforced by the adoption of milking robots. Since cows return several times a day to the robot for milking, the grazing area that can actually be grazed significantly reduced as it is limited by the distance to the robot. Cows may thus need to continuous supplementation of green forage.

Forage conservation

Since the 1980s, round or square bale silage was very successful in specialised grassland regions. Notwithstanding its higher cost compared to clamp silage, this system was largely adopted because it is very flexible, especially in regions were plots are small, the grassland surface of the farm is broken up and the climate is rainy. Big bale silage is typically produced from drier and slightly more mature grass than clamp silage. It is thus well adapted for conserving haylage for suckling cows.

Hay continues to be made in relatively small amounts (< 25% of harvested grassland forage) to provide a fibrous feed to particular classes of stock, e.g. young stock, horses and sheep. It is also produced to align with agro-environmental measures, where species-rich grasslands and field margins must be cut late to improve vegetation diversity, allow reseeding of many flowering plants, and wildlife breeding.

Silage making. Left: Cutting. Right: Wilting in windrows. Source: ILVO - Merelbeke.

Grass harvesting for silage. Left: Demonstration of machinery types. Right: Self-propelled forage harvester filling a wagon pulled by tractor. Source: ILVO – Merelbeke.

Haylage. Left: Round bale protected by a plastic film. Source: ILVO – Merelbeke. Right: Belgian Blue cows eating haylage in a barn in winter. Source: Wallonie Elevages.

Environmental objectives

A policy response began at European level with the adoption in 1992 of regulations on the agri-environmental scheme (EEC 2078/92 and EC 1257/99 regulations), which includes support for organic farming; of the Nitrates Directive (Directive 91/676/EEC) in 1991; and of Natura 2000 (Birds (1979) and Habitats (1992) directives). The agri-environmental scheme was transposed into Belgian regional laws relatively quickly in the 1990s. The Nitrates, Birds and Habitats directives were only fully transposed in the early 2000s.

Agri-environmental programmes are designed at regional level in Belgium. The adoption of this programme and the associated payments are optional for farmers. They have two main objectives: reducing environmental risks associated with intensive farming and preserving biodiversity and landscapes. Agri-environment payments may only be made for actions above the reference level of mandatory requirements defined by codes of good farming practices (GFP). Agri-environmental measures (AEM) include the support of the conversion to and maintenance of organic farming.

▸▸ Opportunities to improve pasture resources

Since the 1990s, the negative environmental impacts of intensive grassland management have become increasingly recognised. Nitrate leaching and water pollution by nitrate and phosphate were the first topics to emerge in grassland research on environmental protection. The best techniques for slurry application were studied in detail. The effects of the total amount and the date of application of mineral and organic fertilisers, as well as the effect of stocking rate and supplementary feeding in grassland paddocks on the nitrate residue in soil profiles in autumn, were investigated. Nutrient balances were calculated at field and farm-gate levels. Research on biodiversity conservation and restoration along with the integration of species-rich grasslands in efficient and profitable stockbreeding systems quickly followed studies on water pollution. New research began on legumes that had been somewhat ignored by grassland research over the previous thirty years. The inclusion of legumes in grassland swards was also seen as a way to increase sward diversity. 'Secondary' or wild grass species were studied within a context of biodiversity restoration, of extensification and to better understand the competition between grasses. In the early 21st century, greenhouse gas emissions from grassland soils, ruminants and animal manure during storage and spreading became important research topics. The positive influence of green forage on the conjugated linoleic acid (CLA) and polyunsaturated fatty acid (PUFA) content of milk and meat received considerable attention for their positive effects on human health. However, holistic approaches developed slowly compared with scientific efforts in reductionist approaches.

Future research must still address these issues by integrating agricultural management with environmental protection. In particular, new research priorities should be related to the four priorities defined in the 'Health Check' set out in the 2008 CAP: climate change, renewable energy, water management and biodiversity. New research should focus on energy consumption of agricultural production, GHG emissions and biodiversity conservation and restoration.

Climate change and energy savings and production issues are intrinsically linked. Less energy use by agriculture will lower CO_2 emissions into the atmosphere. Future increases in fossil fuel prices and the resulting rise in nitrogen fertiliser prices will be strong driving forces of change in Belgian farming systems over the next decades. Fundamental system reforms are inevitable. In the future, systems will have to be less energy demanding. Fortunately, there are important possibilities to save energy per ha or per tonne of milk, for instance (Haas *et al.*, 2001). Future research should systematically focus on energy costs and GHG emissions per production system and per product to develop such energy-efficient systems. A new integration of legume-based grasslands in crop rotations of arable land at the farm and/or the region levels will be necessary to restore arable soil fertility, store carbon in soils, use less nitrogen fertiliser, and reduce transportation costs. Biological nitrogen fixation by legumes will be one of the pillars of these future systems in order to save the huge amounts of fossil energy required for the synthesis of nitrogen fertiliser.

Climate change predictions by the Intergovernmental Panel on Climate Change (IPPC) suggest higher mean temperatures and changes in rainfall distribution (more summer droughts and heavier winter rain) in Belgium. Adaptation measures should be adopted. *Dactylis glomerata* and *Festuca arundinacea* could, for instance, replace *Lolium* spp. in cutting and mixed-used (cutting and grazing) systems. New management systems should be designed to determine the best techniques for using *D. glomerata* in mixed-used swards since this grass cannot hold up to intensive grazing. Legumes and especially *Medicago sativa* and *Trifolium pratense* will certainly have an advantage over grasses in summer drought periods. Systems will have to change by starting grazing earlier in spring, later grazing in autumn and higher silage use in summer. Nitrogen fertilisation patterns will also need to be adapted. Since climate change models are predicting more brutal rainy precipitation levels in the future, grasslands could also be used for flood buffering and enhancing water infiltration. This objective could be combined with wetland and biodiversity restoration. Research could examine the optimisation of these changes over time as they progressively become more common.

Climate change could also be mitigated in grasslands through carbon storage. This can be achieved by converting arable land to grassland with better crop/livestock integration. Biomass produced by lucerne grown for energy could possibly contribute to biofuel production on arable land. Hedge planting and woodland strip plantation on grassland margins are also options. Research should study the potential benefits of these techniques.

There are possibilities to increase the role of many of the lesser-used grassland species (grasses and legumes) for particular environments and species-rich sward restoration. This is especially applicable to upper Belgium. Grassland species can be used for several environmental enhancement purposes like grassy field margins for erosion control in arable land, surface water protection (buffer zone) and water infiltration in grasslands and arable land, species-rich field margins for restoring farmland bird and pollinator populations, for example.

Grasslands should produce high amounts of quality water by favouring clean water infiltration. Nitrate residues in soil water should thus be reduced to a minimum while water infiltration should be enhanced. Trampling of grassland soils by high animal stocking rates and heavy grass harvesting machines for silage making have often degraded soil structure, in turn causing poor water infiltration in grassland soils. Future research should examine ways to improve this situation. Continuous improvement of nutrient management should be a target of future research to reduce production costs and pollutions.

Grassland forages are now recognised for having a beneficial effect on animal product composition and human health. Conjugated linoleic acid (CLA) content in meat and dairy products can have a positive anti-carcinogenic effect while a low ratio of n-6 to n-3 polyunsaturated fatty acids (PUFA) has been associated with a low susceptibility to coronary heart disease. These findings create possibilities for further research on grassland-based feeding systems. Grass-based bull and ox fattening systems should be studied to increase efficiency, profitability and product quality. Similar research should be conducted in dairy systems for dairy product quality. Grassland-based systems of tasty products should be developed in dairy, beef meat, pig and poultry productions. New livestock breeds could be introduced (e.g. Aberdeen Angus and rustic pig breeds) or created by selection with a view to better use of grass and quality product diversification. New quality production systems should be developed or improved while minimising their possible negative impacts on the environment, or which combine both biodiversity restoration and the production of quality animal products.

Future grassland and grassland-based system research should focus on the multiple functions that grasslands offer society. Forage and livestock systems should be supported by society for the services they provide, including biodiversity conservation and landscape protection. They also offer recreational opportunities and contribute to quality of life by producing healthy and tasty products. Future research should take all these aspects into account.

Industrial livestock systems should be fundamentally reformed for many reasons: poor animal welfare, low meat quality, negative impacts on the environment and biodiversity, and dependence on fossil fuel. Future systems need to release fewer GHGs into the atmosphere (CO_2 as well as CH_4 and N_2O) and have a much better impact on biodiversity.

Agroecological systems that reduce costs and raise revenue should be urgently developed. Revenue is increased in these systems by targeting quality products and processing and selling them in short and local marketing chains whenever possible to use local resources instead of commercial inputs. Such systems replace fossil-fuel based inputs through the ecosystem services provided by biodiversity (e.g. biological nitrogen fixation, pollination, pest control by natural enemies). They are adapted to climate change and have the potential to mitigate climate change by storing carbon in soils and vegetation. They are a credible way to address the future increase of input prices caused by fossil fuel rarity and are an opportunity to increase product quality expected by consumers.

Policy research should examine the efficiency and reform of existing programmes: reinforcement of the cross-compliance principle, agro-environmental schemes, support for organic farming, implementation and development of Natura 2000, implementation of the High Nature Value (HNV) farming programme, the quality product policy, and the integration of ecosystem services provided by grasslands in the price of products or financial support. Agriculture and food policies should support the conversion to agroecology. Research should thus look at the design of new policy programmes to achieve the multiple objectives of future Belgian agriculture.

▸▸ References

Bondesen O.B., 2006. The global seed production now and in the future, where and who? Danish Seed Council. Available at: http://www.dansklandbrug.dk/NR/rdonlyres/F07E001C-702E-45B1-A66E-32BC6CF338E0/0/Global_seed_now_and_future.pdf

Carels K., De Clercq P., Van Gijseghem D., 2005. Impacts of Agricultural Policy on Rural Development in Belgium: case study of the Flemish Region. OECD workshop on Evaluating Agri-environmental Policies, Bratislava, 24-26 October 2005: 13 pp.

CBD-Belgium, 2009. Fourth National Report of Belgium to the Convention on Biological Diversity. Royal Belgian Institute of Natural Sciences, 96 pp.

Deprez B., Lambert R., Decamps C., Peeters A., 2004a. Production and quality of red clover (*Trifolium pratense*) and lucerne (*Medicago sativa*) in pure stand or in grass mixture in Belgium. Grassland Science in Europe, 9: 498–500.

Deprez B., Lambert R., Decamps C., Peeters A., 2004b. Nitrogen fixation by red clover (*Trifolium pratense*) and lucerne (*Medicago sativa*) in Belgium leys. Grassland Science in Europe, 9: 469–471.

Deprez B., Parmentier R., Lambert R., Peeters A., 2007. Les prairies temporaires : une culture durable pour les exploitations mixtes de la Moyenne-Belgique. Les Dossiers de la Recherche agricole 2. Namur, Belgium, Ministère de la Région Wallonne, Direction Générale de l'Agriculture, Direction de la Recherche: 84 p.

DGSEI, 1990. L'agriculture. Available at: http://statbel.fgov.be/fr/statistiques/chiffres/economie/agriculture/index.jsp

DGSEI, 2000. L'agriculture. Available at: http://statbel.fgov.be/fr/statistiques/chiffres/economie/agriculture/index.jsp

DGSEI, 2008. L'agriculture. Available at: http://statbel.fgov.be/fr/statistiques/chiffres/economie/agriculture/index.jsp

Federal Public Service Economy (FPS Economy), 2007. Panorama de l'économie belge. Brussels. 257 pp.

FPS Economy – DGSEI, 2008. L'accroissement de population le plus important depuis 1965. La Belgique s'approche des 11 millions. Communiqué de presse, 28 août 2008 : 3 pp.

FPS Economy – DGSEI, 2009. Population. Available at: http://statbel.fgov.be/fr/statistiques/chiffres/population/index.jsp

FPS Economy, 2008. L'agriculture en Belgique en chiffres. Chiffres clés de l'agriculture belge 2008. Directorate-General Statistics and Economic Information. Brussels. 24 p.

Genot ed., 2005. Les ressources génétiques des animaux d'élevage en Belgique. Contribution de la Belgique au Premier Rapport sur l'État des Ressources Zoogénétiques dans le Monde. Rapport national à la FAO. 58 pp. Available at: agriculture.wallonie.be/apps/spip_wolwin/IMG/pdf/RapportNationalFAO.pdf

Haas G., Wetterich F., Köpke U., 2001. Comparing intensive, extensified and organic grassland farming in southern Germany by process life cycle assessment. *Agriculture, Ecosystems and Environment*, 83, 43–53.

Hanset R., 1998. Emergence and selection of the Belgian Blue breed. Belgian Blue Herd-Book. Ciney, Belgium, University of Liege. 16 pp.

ISF (International Seed Federation), 2007. Certified (C) and uncertified (UC) seed production (tons) of selected forage and turf species. Sowing season 2006. Available at: http://www.worldseed.org/cms/medias/file/ResourceCenter/SeedStatistics/ForageandTurfSeedMarket/

Natagriwal, 2015. Dossier de presse. 1995 – 2015: 20 années de Mesures Agro-Environnementales (MAE) en Wallonie. Natagriwal: 5 pp.

Organisation for Economic Co-operation and Development (OECD), 2008. Environmental Performance of Agriculture in OECD countries since 1990. Belgium Country Section, pp. 210-242. Paris.

Peeters A., Kopec S., 1996. Production and productivity of cutting grasslands in temperate climates of Europe. Grassland Science in Europe, 1: 59–73.

Peeters A., Parente G., Le Gall A., 2006. Temperate legumes: key species for sustainable temperate mixtures. Grassland Science in Europe, 11: 205–220.

Statbel, 2008. https://statbel.fgov.be/fr.

Wong D., 2005. World Forage, Turf and Legume Seed Markets. Available at: www1.agric.gov.ab.ca/%24department/newslett.nsf/pdf/fsu6885/%24file/worldforage.pdf

Germany

Lena Dangers and Felicitas Kaemena

▸▸ Introduction

Animal production: facts about dairy farming in Germany

Germany is the biggest milk producer in the EU and milk production is one of the most important sectors of German agriculture. It accounts for 19% of the production value of German agriculture and is the main source of income for a quarter of all agricultural farms in Germany.

The structure of dairy farming has changed in recent years in Germany. The number of dairy farms has continuously fallen while the number of dairy cows has not changed (Figure DE1), which means that the number of cows per farm is rising (Thünen Institute, 2018). From 1970 to 2017, the number of farms decreased from 838,000 to 66,000 dairy farms. The average number of dairy cows per farm rose from 7.3 to 64 cows per farm.

Furthermore, the amount of cow milk produced has risen slightly. In 2009, 29.2 m tonnes of cow milk were produced in Germany. In comparison, 31.3 m tonnes of cow milk were produced in 2017 (BMEL, 2017).

Reviewer: University of Göttingen.

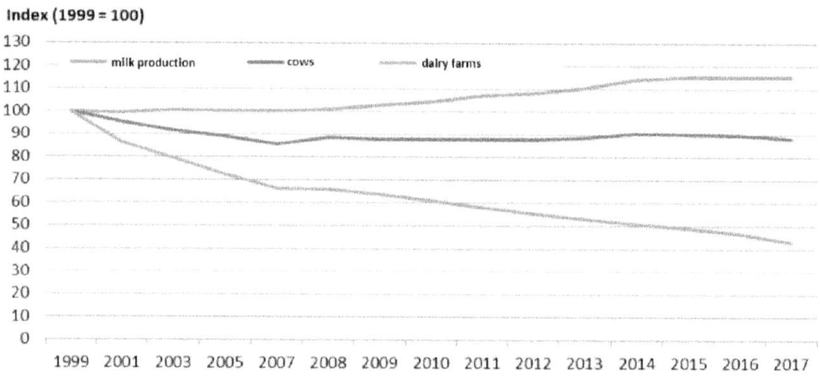

Figure DE1. Development of milk production, number of cows and dairy farms from 1999 – 2017 (Thünen Institut, 2018).

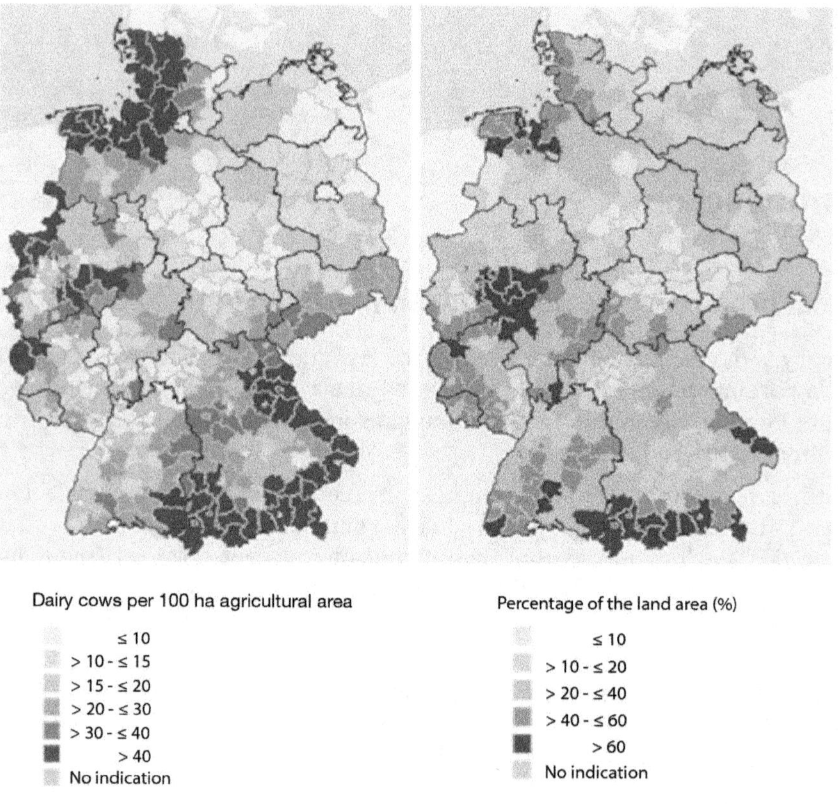

Figure DE2. Relation of number of dairy cows and the percentage of grassland of area used for agriculture (Thünen Institut, 2018).

Figure DE2 shows that the milk production is strongly concentrated in some regions. Almost half of all dairy cows in Germany are in Bavaria and Lower Saxony. This is because dairy farms are mainly located in grassland regions. Grassland is a valuable feed for ruminants and explains the regional distribution of milk production (Thünen Institut, 2018). The milk quota system was simplified from 2007 to 2015, with milk quotas able to be traded in bigger areas from west to east and vice versa. Previously, it was only possible to trade within smaller regions. This change led to an increase in milk production in north-western Germany.

The average herd size has increased in the last ten years in nearly all federal states. Very large herds are located in eastern Germany due to historical reasons. The average herd size in eastern federal states is 188 cows, while in the other federal states, an average of 54 dairy cows are kept per farm. In a nationwide comparison, Bavaria has the smallest dairy farms with an average of 37 cows per farm (Table DE1).

Table DE1. Average herd size in Germany (own table, based on Thünen Institut, 2018).

Federal state	Average herd size (cows per farm)
Baden-Württemberg	42
Bavaria	37
Brandenburg	226
Hesse	47
Mecklenburg-Western Pomerania	224
Lower Saxony	84
Nordrhine-Westphalia	66
Rhineland-Palatinate	59
Saarland	69
Saxony	206
Saxony-Anhalt	143
Schleswig-Holstein	93
Thuringia	175

German dairy farming is predominately characterised by indoor systems. About 58% of dairy cows in Germany are kept in a confined system (full-year-housing) (Gurrath, 2011). Input levels input differ from farm to farm. There are high, moderate and low input systems. Due to climate conditions, farm structure and other aspects, dairy cows are kept indoors for at least six months of the year. In 2010, 70% of dairy cows were kept in modern freestall barns with bedded cubicles and a long feed table (BLE and Bundesinformationszentrum Landwirtschaft, 2019). As new buildings are in most cases freestall barns, it is assumed that the figure of 70% of dairy cows in freestall barns has even increased since 2010. An advantage of modern housing systems is the spatial separation of eating, laying and milking areas. Furthermore, they provides cows with a comfortable bed, protection from the elements and free access to a well-balanced diet. It is also possible to store and manage slurry and bedding material.

Tie-stalls continue to be a common practice on a few dairy farms. These farms usually have smaller herds and are located in southern regions, such as the Alps. According to the Association of German Cattle Breeders (ADR), the number of tie-stall systems in Germany fell by 77% between 1995 and 2013 (BLE and Bundesinformationszentrum Landwirtschaft, 2019).

Moreover, automatic milking systems (AMS) have become more popular on German dairy farms. In 2016, around 7,800 AMS were used on 5,500 farms in Germany. Two thirds of all dairy farmers who decide to invest in their farms buy a milking robot. Besides the AMS, on many farms automatic feeding systems and electronic transponder feeder systems are used (BLE and Bundesinformationszentrum Landwirtschaft, 2019)

There are 4.2 million dairy cows in Germany, of which two-thirds are pure dairy breeds. Most of the dairy cows are Holstein Friesian (HF) cows (60%). In some regions other breeds like Fleckvieh, Braunvieh or crossbreeds are more common. They are dual-purpose breeds used for milk and meat production. Around 31% of dairy cows are Fleckvieh, and this is the dominant breed in Bavaria and Baden-Württemberg. In the northern and eastern federal states, black/white and red/white HF are the predominant breeds (Statistisches Bundesamt, 2015; Lindena and Lassen, 2016).

Dual-purpose breeds have a lower milk yield but produce more meat than pure breeds. The average milk yield in Germany is 7.746 kg/cow/year (BLE and Bundesinformationszentrum Landwirtschaft, 2019).

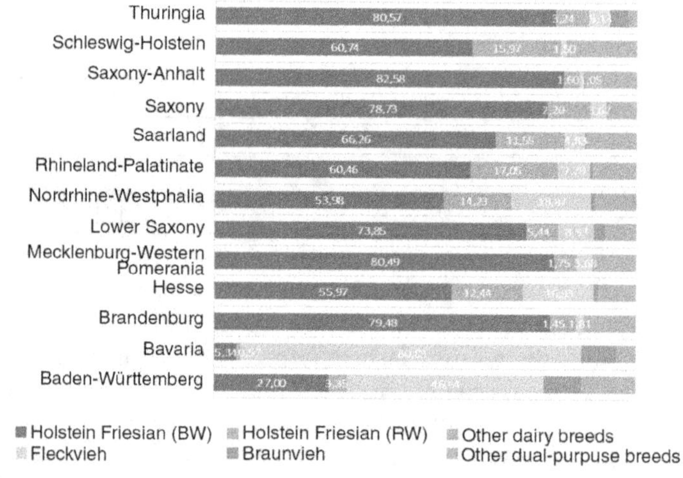

Figure DE3. Distribution of cow breeds in Germany (own figure, based on Statistisches Bundesamt 2015, Land- und Forstwirtschaft, Fischerei – Viehbestand, Fachserie 3 Reihe 4.1).

Importance of grassland in Germany

Grassland has an important role in Germany and is characteristic in many regions. In 2017, about 4.7 million hectares were used as permanent grassland. Grasslands in the country are typically used in a meadow system (39% of permanent grasslands) and 56.5% of permanent grasslands is grazed (Statistisches Bundesamt, 2019). Due to a growing demand for high-energy feed and plants for renewable energies, the proportion of permanent grasslands has decreased by 5% since 2003. Former grassland areas were converted to arable land, and even sensitive sites such as Natura 2000 protected areas, peatland, charted habitats and river floodplains were ploughed. The conversion causes problems for nature and climate and results in a loss of diversity (BfN, 2018).

Table DE2. Different uses of permanent grassland areas (Statistisches Bundesamt, 2019).

	Acreage in 1000 ha					
	2010	**2013**	**2014**	**2015**	**2016**	**2017**
Permanent grassland	4654.7	4621	4650.7	4677.1	4694.5	4715
Meadows (average use)	1899.2	1826.8	1829.6	1844	1876.8	1843.3
Grazing (including mowed pastures and mountain pastures)	2544.7	2584.6	2620.3	2651	2630.6	2664.4
Low-yield permanent grassland	188	191	183.2	164.9	170	187.3
Permanent pasture (no longer in production) with entitlement to aid/ premium	22.8	18.6	17.5	17.2	17.1	19.9

▶▶ Silage

The main feed on German farms is fermented feeds such as maize and grass silage, which is generally produced on farmland. Grass silage is predominately used as a feed source on dairy farms in Germany. Grass silage is important in feed structure and as a source of protein. In recent years maize silage has become more important as an energy feed source. This is due to the fact that the milk yield per cow increased and the demands for consistent feed quality became higher. German farmers need to focus on high-quality silage, because dairy cows are indoors for at least six months per year and calve throughout the year. This sets the German system apart from the grass-based pasture systems in New Zealand or Ireland.

Concentrates are another important feed source used to maintain a high milk yield and a good nutrient supply for high-yielding cows.

Table DE3. Targets for high-quality maize and grass silage (own table based on DLG, 2006).

	Grass silage	Maize silage
DM %	30 - 40	28 - 35
Ash % of DM	< 10	< 4.5
Crude fiber % of DM	22 - 25	17 - 20
NDForg % of DM	40 - 48	35 - 40
Starch % of DM	-	> 30
NEL MJ NEL/kg DM	> 6.4 or > 6.0	> 6.5
Crude protein g/kg DM	> 135	> 130

▸▸ Fermentation basics of silage making

The fermentation of silage is strongly influenced by lactic acid fermentation, dry matter content (osmotic pressure), pH and nitrate content. In many but not all cases, successful fermentation can be explained by various biological factors such as dry-matter content, buffering capacity and sugar content. Furthermore, management factors such as silo filling speed, compacting, type of additives used, chopping length and silo management can also affect the fermentation of grass or maize. In some cases, a poor fermentation process can explain poor silage quality with a low nutritive value and a low feed intake (Kung and Shaver 2001). Therefore, German farmers take usually samples of each silage pit to evaluate fermentation success and feed quality. Based on the fermentation analyses, farmers determine the feeding management.

▸▸ Anaerobic conditions

Anaerobic conditions are very important in the first stage of fermentation. The following points should be kept in mind:
– Anaerobic conditions can be created by compacting and storing the plant material under a sealed area.
– Oxygen cannot be removed through compacting; it is rapidly removed by plant respiratory processes.
– Sealing prevents escaping of CO_2 and the re-entry of air during storage.
– Any contact between the plant material and air initiates aerobic deterioration.
– The activity of aerobic microorganisms can result in decayed, inedible and sometimes toxic material.
– Aerobic deterioration increases dry matter losses and reduces nutritional value.

Buffering capacity (BC) measures the degree to which a forage sample will resist a change in pH. All types of forage have different buffering capacities. Fresh forage with a high buffering capacity requires more acid to reduce its pH than forage with a low buffering capacity. In general, fresh legumes have a higher buffering capacity than fresh grasses or maize (Table DE4).

Table DE4. Fermentability of forage (DM%: dry matter; WSC: water soluble carbohydrates; BC: buffering capacity; LA: lactic acid) (LWK Niedersachsen).

Plants	DM %	WSC in g/kg DM	BC in g LA/kg DM	WSC/BC-quotient
Whole crop maize (milk maturity)	22	230	35	6.6
Whole crop maize (paste maturity)	30	110	32	3.4
Ryegrass - fresh	20	173	52	3.3
Ryegrass - wilted	35	173	52	3.3
Other grasses - fresh	20	92	55	1.7
Other grasses -wilted	35	92	55	1.7
Red clover - fresh	20	115	69	1.7
Red clover - wilted	35	115	69	1.7
Alfalfa - fresh	20	65	74	0.9
Alfalfa - wilted	35	65	74	0.9

Buffering capacity
— Amount of lactic acid which is necessary to lower the pH value to < 4.0
— Resistance of plant material to adverse pH shift
— Raw protein, minerals and soil materials have buffering properties

Extremely wet silages (< 25% DM), prolonged fermentation (due to high buffering capacity), loose packing or slow silo filling can result in silages with high concentrations of acetic acid (> 3 to 4% of DM). In such silages, energy and DM recovery are probably not ideal. Silages treated with ammonia also tend to have higher concentrations of acetic acid than untreated silage, because fermentation is prolonged by the addition of the ammonia which raises the pH (Kung and Shaver 2001).

High concentrations of ammonia (>12 to 15 % of CP) are the result of excessive protein breakdown in the silo caused by a slow drop in pH or clostridial action. In general, wet silages have higher concentrations of ammonia. Extremely wet silages (< 30 % DM) have even higher ammonia concentrations, because of the potential for clostridial fermentation. Silages packed too loosely and filled too slowly also tend to have high ammonia concentrations (Kung and Shaver 2001).

▸▸ Grass types

Perennial ryegrass is the most common grass type used on German grasslands for silage production. Due to high yield potential and good feed quality it is characterised by a relative high amount of water soluble carbohydrates (WSC, sugar) compared to other grass types (Table DE5, LWK).

Table DE5. Types of grasses and the fermentability (DM%: dry matter; WSC: water soluble carbohydrates; BC: buffering capacity; LA: lactic acid) (LWK Niedersachsen).

Grass types	DM in %	WSC in g/kg DM	BC in g LA/kg DM	WSC/BC-quotient
Italian ryegrass, WV	20.0	190.0	55	3.5
Perennial ryegrass, WD	21.0	155.0	44	3.5
Timothy, WLG	22.0	75.0	40	1.9
Orchard grass, KG	22.0	95.0	43	2.
Meadow fescue, WSC	23.0	90.0	55	1.6
Bluegrass, WRP	19.0	80.0	53	1.5

Further, the right cutting date has a strong influence on the feed quality. Figure DE4 shows the varying energy content (EC) of different grass types at different development stages.

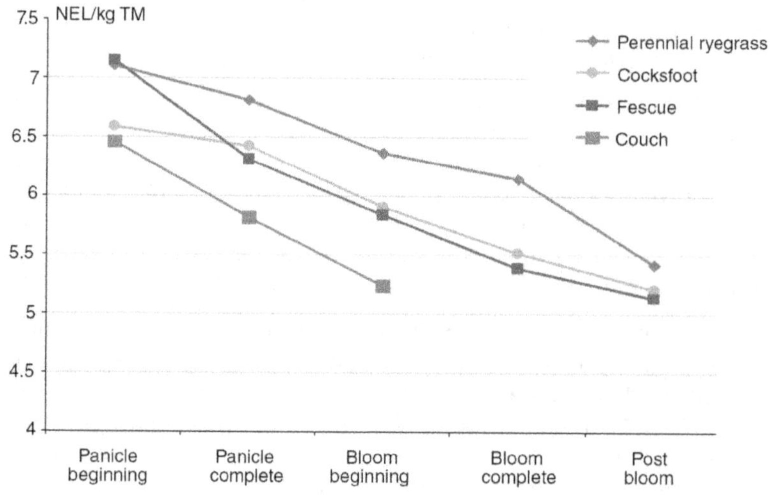

Figure DE4. Relation between the energy content (EC) and the development stages of different grass types (LWK Niedersachsen).

▸▸ Grazing

Grazing during the summer months is still a common practice on many German dairy farms. About 42% of dairy cows graze for five months per year on average (BLE, 2019). However, providing access to pasture and ensuring an active uptake of fresh grass is perceived differently. Some farms concentrate on the advantages with regard to cow health, animal comfort, and environmental and economic aspects, while others include fresh grass as an important component of their cows' diet.

However, there are limiting factors to providing access to pasture in some regions (Pries, 2004). Due to structural changes on German dairy farms, it seems difficult to provide enough grazing area for larger herds with more than 200 cows. Overly small paddocks and long distances between paddock and the milking parlour limit the grazing time and feed intake per cow from fresh grass. Medium-sized farms offer more time on pasture than farms with large herds. Every second cow at medium-sized farms (50 to 200 cows) has access to pasture (Figure DE5).

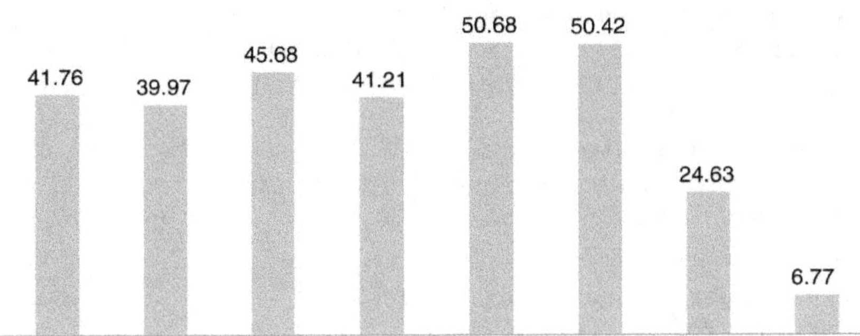

Figure DE5. Relation between herd size and number of dairy cows on pasture in Germany (own figure based on Destatis, 2010).

In German grassland regions it is more common to give cows access to pasture. In the federal states of North Rhine-Westphalia, Schleswig Holstein and Lower Saxony more than 50% of dairy cows have access to pasture (Figure DE6).

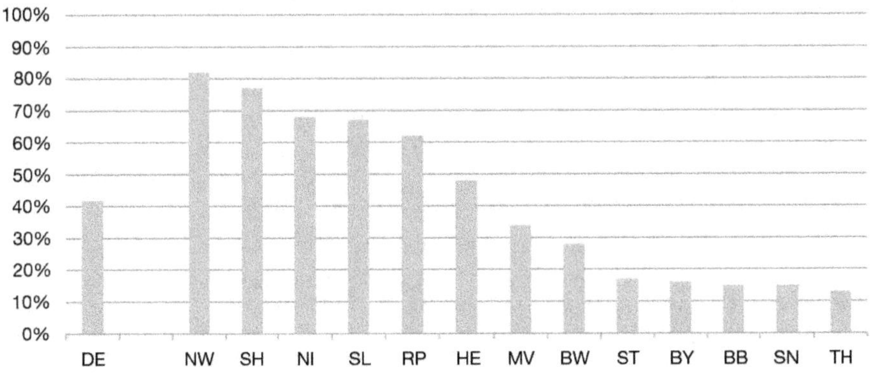

Figure DE6. Percentage of grazing in Germany (DE) and the individual federal states (Thünen Institut, 2018).

There are various grazing system in practice in Germany. The overall aim of the different grazing systems is to obtain as much milk as possible directly from grass. It is also important to ensure constant feed quality and minimal feed losses. The choice of grazing system is independent of actual grazing time.

Short sward grazing (*Kurzrasenweide*) (Steinberger, 2011; Steinberger, 2017)
– Allocation of the entire paddock
– Pre-grazing height 5 to 6 cm
– Residual height 4 to 3.5 cm
– Frequent bites limits dry matter absorption
– High-quality feed, high digestibility
– Good palatability of the grass
– Few feed losses
– More milk per ha
– Dense sward due to short rotation length

Rotational grazing (Grünlandzentrum, 2018)
– Pre-grazing height 8 to 11 cm; graze swards at three-leaf stage
– Post-grazing height 3.5 to 4 cm; depends on growth rate
– High dry matter yield
– Grass offer is adapted to cow demand
– Removable fences offer access to only a part of the paddock
– 2 to 4 grazings/paddock
– Longer resting time than grazing time

Continuous grazing (Grünlandzentrum, 2018)
– Allocation of a pasture block
– Block size at least 4 ha
– Graze block for 3 to 6 weeks
– Labour-saving

▸▸ References

BfN, Federal Agency for Nature Conservation, 2018. Grassland conservation in Germany. https://www.bfn.de/en/activities/agriculture/grassland-conservation-in-germany.html

Bundesanstalt für Landwirtschaft und Ernährung (BLE), Bundesinformationszentrum Landwirtschaft, 2019. Milchviehhaltung in Deutschland https://www.praxis-agrar.de/tier/rinder/milchviehhaltung-in-deutschland/ (aufgerufen am 07.01.2019)

DLG, 2006. Praxishandbuch Futterkonservierung. DLG Verlag, 7. Auflage

Eurostat, 2017. Agriculture, forestry and fishery statistics, p. 85 https://ec.europa.eu/eurostat/documents/3217494/8538823/KS-FK-17-001-EN-N.pdf/c7957b31-be5c-4260-8f61-988b9c7f2316

Gurrath P., 2011. Landwirtschaft auf einen Blick. Wiesbaden, Germany.

Grünlandzentrum Niedersachsen Bremen e.V, 2018. Projektergebnisse „Maximierung der Weideeffizienz" und Projektergebnisse Weidecoach.

Kung L., Shaver R., 2001. Interpretation and Use of Silage Fermentation Analysis Reports. Focus on Forage, 3 (13), 2001.

https://fyi.uwex.edu/forage/files/2014/01/Fermentation.pdf

Lasar A., 2017. Klimaeffizienz von Milchviehhaltungsverfahren; Landwirtschaftskammer Niedersachsen.

Lindena T., Lassen B., 2016. Development, strategies and challenges in the German dairy sector. http://www.eaap.org/Annual_Meeting/2016_belfast/S02_09_Lindena.pdf

LWK Niedersachsen: Fachbereich Grünland und Futterbau

Pries M., 2004. Weidegang ja – aber richtig ergänzen. https://www.landwirtschaftskammer.de/landwirtschaft/tierproduktion/rinderhaltung/fuetterung/weidegang-ja.htm

Statistisches Bundesamt, Destatis, 2010. Landwirtschaftszählung 2010. Weidehaltung von Milchkühen auf Betriebsflächen nach Bestandsgrößenklassen und Bundesländern 2009 https://www.destatis.de/DE/ZahlenFakten/Wirtschaftsbereiche/LandForstwirtschaftFischerei/Landwirtschaftszaehlung2010/Tabellen/9_4_WeidehaltungMilchkuehe.html

Statistisches Bundesamt, 2015. Land- und Forstwirtschaft, Fischerei – Viehbestand. Fachserie 3 Reihe 4.1 https://www.destatis.de/DE/Publikationen/Thematisch/LandForstwirtschaft/Viehbestand/ViehbestandTierischeErzeugung/Viehbestand2030410155324.pdf?__blob=publicationFile

Statistisches Bundesamt, 2019. Landwirtschaftlich genutzte Fläche: über ein Viertel ist Dauergrünland. https://www.destatis.de/DE/ZahlenFakten/Wirtschaftsbereiche/LandForstwirtschaftFischerei/FeldfruechteGruenland/AktuellGruenland2.html

Statistisches Bundesamt, 2019. Feldfrüchte und Grünland. Dauergrünland nach Art der Nutzung im Zeitvergleich

https://www.destatis.de/DE/ZahlenFakten/Wirtschaftsbereiche/LandForstwirtschaftFischerei/FeldfruechteGruenland/Tabellen/ZeitreiheDauergruenlandNachNutzung.html

Steinberger S., 2011. Kurzrasenweide – der Weideprofi misst seinen Grasaufwuchs. https://www.lfl.bayern.de/mam/cms07/ite/dateien/31061_anleitung_zur_grasaufwuchsmessung.pdf

Steinberger S., 2017. Die Weideprofis starten jetzt durch. BLW 10, 10.03.2017, p.32

Thünen Institut, 2018. Steckbriefe zur Tierhaltung in Deutschland: Milchkühe https://www.thuenen.de/media/ti-themenfelder/Nutztierhaltung_und_Aquakultur/Haltungsverfahren_in_Deutschland/Milchviehhaltung/Steckbrief_Milchkuehe2018_final_2.pdf

Poland

Piotr Goliński and Barbara Golińska

▸▸ Introduction

Grassland status

Poland is a country with varied topographic, climatic and edaphic conditions. This diversity has multiple consequences, which manifest primarily in land use. Permanent grasslands occupy a total area of 3.2 million hectares in Poland (Table PL1), which constitutes 21.7% of the total utilised agricultural area or 10% of all of the country's land. The above-mentioned area comprises natural, semi-natural (i.e. periodically subjected to renovation) and agriculturally improved permanent grasslands (renovated, fertilised and intensively utilised for forage production). In regions with a limited area of permanent grasslands and a predominance of arable land, temporary grasslands play an important role in the production of bulky fodder. Grass-legume mixtures are particularly important among plants cultivated on such grasslands because their production potential and high nutritional value ensure that the feeding requirements of high-production ruminants are met. Nowadays, leys established on arable land covering about 0.4 million ha, with a maximum utilisation time of four to five years (Table PL1).

Table PL1. Changes in the utilised agricultural area (UAA) in Poland (million ha) (GUS, 2018).

Type of land	1990	2000	2005	2010	2017
Arable land	14.388	13.940	12.220	10.428	10.757
of which Temporary grasslands	n/a	n/a	n/a	n/a	0.414

Type of land	1990	2000	2005	2010	2017
Permanent meadows	2.475	2.503	2.528	2.629	2.796
Permanent pastures	1.585	1.369	0.859	0.654	0.375

In recent decades, the proportion of permanent grassland in Poland has decreased from 4.06 (1990) to 3.17 (2017) million ha (Table PL2). In comparison with the state of grasslands in 1990, the share of permanent meadows in UAA increased by around 40%, but the percentage of permanent pastures in UAA has fallen by almost 70%. In 2017 meadows made up 88% and pastures 12%.

Table PL2. Changes in the area and structure of permanent grassland in Poland (GUS, 2018).

Permanent grassland	1990	2000	2005	2010	2017
Area (million ha)	4.06	3.87	3.39	3.28	3.17
Percentage of UAA including:	22.0	21.7	21.3	21.8	21.7
Meadows	13.4	14.1	15.9	17.4	19.1
Pastures	8.6	7.6	5.4	4.4	2.6

Figure PL1. Location of permanent grassland in Poland based on remote sensing (Dąbrowska-Zielińska *et al.,* 2015).

The distribution of grasslands in Poland varies with their higher proportion in the structure of UAA in north-eastern Poland as well in the south-eastern part of the country (Figure PL1). Most grasslands are situated in river valleys and mid-field depressions as well as in foothill and mountain areas, where it is impossible to grow ordinary crops. Water is the main factor determining the properties of grassland habitats and phytocoenoses in Poland (Grzyb and Prończuk, 1995). Achieving sufficient humidity in grassland habitats depends on adequate water management, particularly with regard to maintaining the operational water and drainage facilities and restoring dysfunctional systems. Action is often required to ensure the collection of any amount of water (small water retention), and it is important to retain water from the catchments of particular rivers in early spring (Nyc and Pokładek, 2008).

The characteristic of Poland's surface features means that 90% of grasslands are covered by lowland meadows. They comprise dry-ground meadows, flooded meadows, boggy meadows and post-boggy meadows. Boggy meadows occur in wet habitats and they are not important for fodder production, but they play an important environmental role. The other types of meadows are used for fodder production. However, insufficient soil humidity levels resulting from small amounts of precipitation lead to decreased persistence and productivity of these meadows, particularly in dry-ground habitats. Insufficient humidity levels in post-boggy meadows lead to the degradation of peat-muck soils occurring there and, as a result, to unfavourable changes in the floristic composition of the phytocoenoses. The optimum environmental and economic effects in valley grasslands (particularly located on organogenic soils) can be achieved by using surface irrigation and upward-irrigation systems (Nyc and Pokładek, 2008). Thus effective water management in a habitat is the basic condition to enable additional measures that aim to improve production from permanent grasslands through fertilisation, management and technology.

Most Polish permanent grasslands have been characterised by great biological diversity (Warda and Kozłowski, 2012). This results from the influence of favourable natural conditions (location in a central part of the continent, in a transitional climate zone, lack of natural barriers in the east and west, varied geological structure and diverse landforms) as well as from the peculiar impact of human activity, which is different to other European countries (uneven degree of industrialisation and urbanisation, low-input farming with a moderate level of fertilisation and land use, as well as extensive and historically permanent forests).

Biodiversity may be very high in permanent meadows. More than 700 species of vascular plants and their presence in 100 different botanical families has been found in grassland communities (Kozłowski and Stypiński, 1997). Some of them are very rare in Europe and they are protected by law. Polish meadows are characterised by the floristic diversity of their sward, which distinguishes them from meadows in other countries, particularly in Western Europe. The main factors determining the floristic diversity of a grassland sward are the properties of grassland habitat and the intensity of farming practice and utilisation. More sward species are also recorded in the plant communities of permanent grasslands occurring on mineral soils than on organic soils (Baryła, 2001).

A proper analysis of the floristic diversity of meadow communities must take into account their syntaxonomy. The floristic diversity of the same associations and communities may vary across different regions of Poland (Kucharski, 1999). Communities in very wet habitats (*Phragmitetea* class) are characterised by a smaller species diversity resulting from a small number of dominant species (Kryszak and Grynia, 2005). A much greater floristic diversity characterises the phytocoenoses of moderately wet and periodically dry habitats of the *Molinio-Arrhenatheretea* class (Warda and Stamirowska-Krzaczek, 2010).

Grazing livestock and their relation to grassland

There is a close relationship between the status of permanent grassland use and the number of livestock for which meadow and pasture swards constitute a feed base. The grazing livestock in Poland reached peak numbers between 1975 and 1989. The maximum cattle population of 10,733,000 head was observed in 1978 and of 4,991,000 head of sheep in 1986. Just after the political regime changes in 1989, the maxima were followed by slow declines until around 1990 to 1991. Between 1991 and 2002, the numbers of cattle and sheep dropped sharply to 5,533,000 and 343,000 head, respectively. After this period, they decreased relatively slowly until 2007. From 2008 the cattle population has grown slowly while sheep numbers are continuing to decrease. According to the latest available statistical data (GUS, 2018), the cattle numbers in 2017 reached a share of 61.1% and with sheep at just 5.7% of the population in 1990 (Table PL3). The numbers of horses between 1990 and 2017 also dropped to around 20% of the 1990 population.

Table PL3. Changes in the number of grazing livestock (thousand heads) in Poland (GUS, 2018).

Farm animals	1990	2000	2005	2010	2017
Cattle	10 049	6 083	5 483	5 742	6 143
of which dairy cows	4 919	3 098	2 755	2 529	2 153
Sheep	4 159	345	316	261	239
Horses	941	550	312	264	185

The difference in the share of permanent grassland in UAA in specific voivodships (highest-level administrative subdivision of Poland, comparable to a province) is related to cattle stock per 100 ha of UAA. As shown in Figure PL2 the highest cattle stock in heads per 100 ha of UAA in 2017 occurred in Podlaskie voivodship. The low density of cattle and sheep in the southern part of Poland based on the high share of permanent grasslands in UAA resulted in their abandonment and impacted the natural afforestation process, which led to losses of open landscapes in this region.

Milk production in Poland in 2017 reached 13,310 million kg (Table PL4) and, in comparison with 1991–1995, it was more than 2000 million kg lower. During the same period of time, annual milk yield per cow increased from 3083 kg to 5687 kg. In the herds under milk recording, which includes around 37% of total dairy cows, the annual milk yield per cow reached the level of 7771 kg.

One of the leading milk production regions in Poland is the Podlaskie voivodship located in the north-eastern part of the country. There are 2500 dairy farms with more than 50 heads in this region; they produce 22% of milk purchased in Poland and 90% of this amount fulfils EU standards. The stocking rate of dairy cows per 100 ha agricultural land in that region in 2018 was 41.6 LSU, while the average stocking rate of dairy cows in Poland was 15.3 LSU (GUS, 2018). For this reason, north-eastern Poland can be described as the region specialising in grassland-based milk production, where their management focusing on sustainable intensification plays an important role.

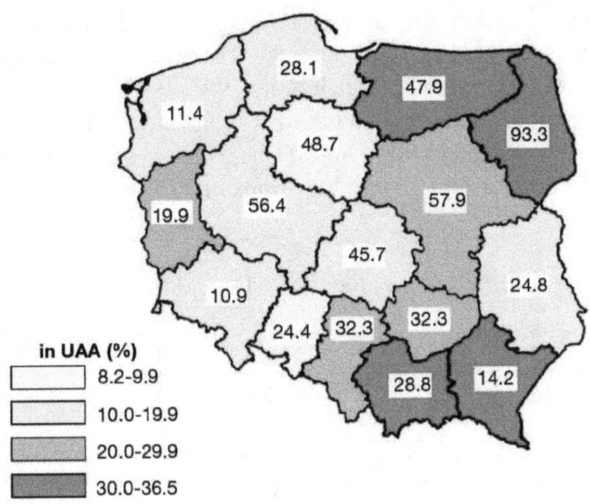

Figure PL2. Share of permanent grasslands in UAA and cattle stock in heads per 100 ha of UAA by specific voivodship (own figure based on GUS, 2018).

Table PL4. Changes in the milk production and yield in Poland (GUS, 2018).

Item	1991-95	2000	2005	2010	2017
Milk production (bn kg)	15.40	11.49	11.58	11.92	13.31
Average annual quantity of milk per cow (kg)	3 083	3 828	4 147	4 487	5 687
Under milk recording (kg)	4 209	5 379	6 508	6 980	7 771

Apart from the Podlaskie voivodship, Mazowieckie and Wielkopolskie are the next largest milk production regions in Poland. They produce 22% and 15%, respectively, of milk purchased in the country. It is worth emphasising that in Mazowieckie, as in Podlaskie, dairy cow feeding is based on feed from temporary and permanent grasslands, while in Wielkopolskie fodder is mainly composed of maize silage.

Nowadays, there are three models of milk production in Poland in terms of grassland fodder use:
– Intensive milk production using a total or partially mixed ration (TMR/PMR) system with an important contribution of silage/haylage from permanent/temporary grasslands (milk yield 10,000 to 12,000 kg per cow), HF breed
– Low-cost technology of milk production with very important role of pasture and/or silage/haylage from permanent/temporary grasslands (milk yield 7000 to 8000 kg per cow), HF breed and/or dairy cattle breeds adapted to grazing
– Organic milk production with crucial role of pasture and/or hay from permanent/temporary grasslands – feeding without silage (milk yield 5000 to 7000 kg per cow), dairy cattle breeds adapted to grazing

From the point of view of market output, which results from the transfer of grassland fodder, grazing livestock meat and wool are also important. A dramatic decline in beef production was observed when comparing the 1991–1995 period and 2005 (Table PL5). Since 2005, a significant increase in beef production with the promising development of beef exports in recent years can be observed. Wool production dropped from 5386 tonnes during the 1991–95 to 620 tonnes in 2010. A slight increase can be seen for 2017.

Table PL5. Changes in beef and wool production in Poland (GUS, 2018).

Type of production	1991-95	2000	2005	2010	2017
Beef production in post-slaughter warm weight (thousand t)	652	321	313	389	563
Wool production – sheep's greasy (t)	5 386	1 322	998	620	774

▸▸ Grassland productivity and management

Permanent grasslands dominate (76.1%) in the structure of fodder area in Poland (Figure PL3). In the analysis, including forage production from main crops, maize cultivation occupies a share of 13.1% and temporary grassland (grass and grass-legume mixtures, as grasses and legumes in pure stands grown for feed) 10.8%.

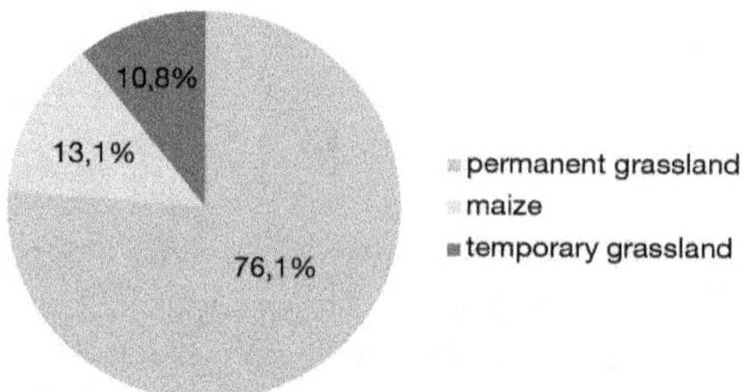

Figure PL3. Structure of fodder area in Poland (own figure based on GUS, 2018).

Because of the low productivity of permanent grasslands in Poland, the feed base for ruminants, particularly for dairy cows, is secured through wilted silage from temporary grasslands and maize silage. As reported by Zastawny (2000), the low-quality fodder from permanent grasslands before 2000 was confirmed by the fact that they occupied 69.9% of the total fodder area in Poland, whereas 30.1% of the area occurred in the form of fodder crops on arable land (including temporary grasslands). However, permanent grasslands produced only 36.0% of nutritional (oats) units in comparison with 64.0% from fodder crops on arable land.

In 2017 the average yield of permanent grasslands in Poland was 5.21 tonnes of hay per hectare (GUS, 2018). The total production of permanent grasslands that year, when taking into account the different yields of meadows and pastures and calculating hay as air-dried matter, was 15.15 thousand tonnes from meadows and

1.37 thousand tonnes from pastures. Although the statistical data show low productivity of grassland, the investigation results regarding determination of DM and crude protein yields carried out on well-managed productive grasslands, particularly on dairy farms, prove that there are a huge forage resources in Polish grasslands. A good example is the synthesis of mutli-year research conducted by Poznan University of Life Sciences (PULS) mainly in Wielkopolska region, where weather conditions due to frequent shortages of precipitation negatively affect the productivity of grassland (Figure PL4).

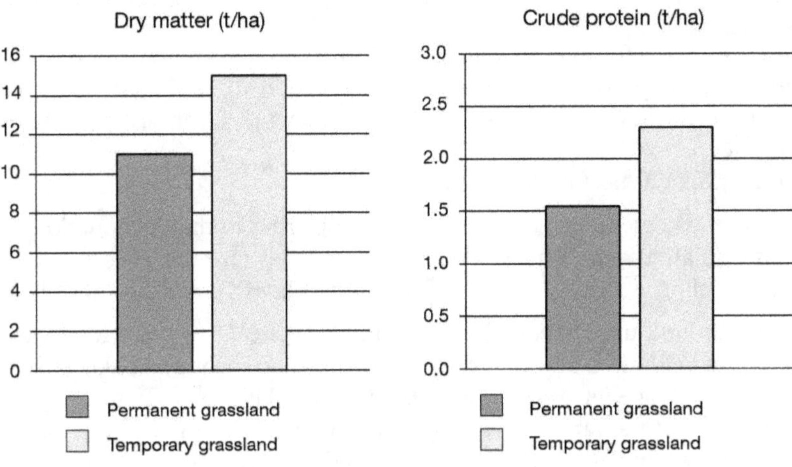

Figure PL4. Potential yields from grassland in Poland (own figure based on synthesis of multi-year PULS research).

The majority of Polish grasslands is utilised by cutting. Dairy cattle grazing has decreased not because of the low effectiveness of this feeding method, but in order to lower labour inputs, particularly on small farms (Wasilewski, 2011). Sward conservation is predominately through hay. An average of 60% of hay production comes from the first cut and about 50% from the second cut. The remaining sward is ensiled, mainly in form of wilted silage containing 30% to 40% of DM. Comparing various regions, the largest amount of silage from meadow swards (30%) is produced in the Podlaskie voivodship (Jankowska-Huflejt and Domański, 2008). Only very little of the harvested sward from grassland (less than 3%) is conserved as dried fodder.

NPK input for the 2015–2016 period was 130.3 kg/ha of agricultural land, of which nitrogen accounts for 71.7 kg/ha, phosphorus 22.4 kg/ha and potassium 36.2 kg/ha (GUS, 2018). It is estimated that the annual input of nutrients in fertilisers on grasslands is smaller and does not exceed (in the case of nitrogen) 50 kg/ha on average.

Following Poland's entry into the EU on 1 May 2004, the country's agricultural situation changed considerably. The following aspects should be noted when grassland management is considered:
– Direct subsidies are calculated per hectare: They are paid on the condition that grasslands are utilised – at least one-time cutting or grazing.

– Financing of programmes connected with the protection of natural environments and rural landscapes: Financial resources are intended to fund various measures in high nature value areas, which, from the EU's viewpoint, possess special significance for natural protection and which are part of the Natura 2000 program. The European Commission has already approved 849 "habitat areas" and 114 "bird areas" in Poland, which cover about 20% of the country. The above-mentioned activities include, among others, maintenance of extensive meadows (one-time and two-time cuttings) as well as extensive pastures (xerothermic plant communities, mountain pastures) and on which renovation – both resowing and complementary seeding – is banned.

– Subsidies intended for milk production modernisation: This includes the purchase of machines for the production, harvesting and conservation of feed, pasture establishment and renovation etc., as well as restructuring of farms specialising in beef cattle and sheep.

– Individual milk quotas (abolished in 2015).

Some of the above elements, especially those referring to milk production, are intended to stimulate the renovation of grasslands and ley farming. This is all the more true since cereal production decreased following Poland's entry into the EU.

The pattern of land use reflects the size and structure of farms. More than 95% of farms are individually- and family-owned. There are around 1,400,000 farms in Poland with an average farm size of 10.8 ha in 2018 (GUS, 2018). In recent years, the structure of agricultural farms has slowly changed. Their average area grew through the purchase or leasing of land, which enhanced their competitiveness and economic efficiency. However, regional differences in farm structures remain. South-eastern Poland is characterised by a large number of small farms, while farms with the largest tracts of land can be found in northern Poland. In 2017, farms over 15 ha in size managed 60.6% of agricultural land, although they constitute only 15.2% of the total number of farms. As of 2016, there were 342,101 farms raising cattle, of which 270,831 were dairy cows (GUS, 2018). For this reporting year, only 9,426 farms had sheep and 9,791 had goats. There were 61,229 horse farms. There were only 11,395 farms where selling farm-based (but not agricultural-based) products and services e.g. agrotourism amounted to more than 50% of the farm's sales.

▸▸ Grassland renovation and establishment of new grass swards

The main causes of grassland degradation in Poland include:

– unfavourable site changes caused either by excess or deficits of water during the vegetation season (e.g. drought in 2018);

– overwintering conditions unfavourable for fodder grass and legume species (the disappearance of *Lolium* sp. coincides most often with freezing of winter barley and rape);

– negligence on the part of farmers with regard to grassland management, especially concerning fertilisation and utilisation;

Re-sowing involving the destruction of the old sward and sowing of a new mixture is recommended in situations of a strong degradation of grasslands which includes:
– very small presence in the sward of valuable grasses and legumes;
– high proportion in the sward (over 40% in coverage) of undesirable and trouble-some rhizomatous (*Cirsium sp.*, *Rumex sp.*), toxic (*Ranunculus sp.*) and hydrophilic (mainly *Juncus sp.*, tussock-forming *Carex sp.*) weeds and mosses;
– over 20% share in the sward of *Deschampsia caespitose*;
– soil impoverishment and a destroyed structure with humic acid content;
– considerable sward destruction by moles, rodents and wild boars.

In the years following the intensified process of meadow degradation, the applied renovation treatments should include complementary seeding and, if production costs of 1 MJ energy continue to increase, the old sward should be destroyed and a new grass-legume mixture sown. It is worth emphasising that complementary seeding should be applied as early as is appropriate.

At the present time, decisions concerning grassland renovations are taken exclusively on an economic basis and depend on the demand for high-quality feed for ruminants. In fact, it mainly applies to farms specialising in dairy cattle production (Figure PL5).

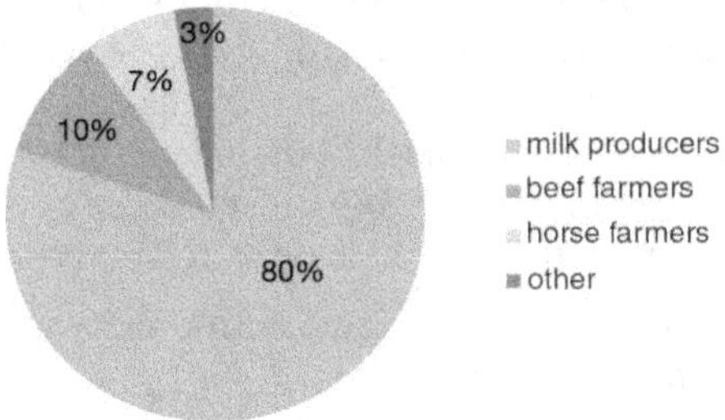

Figure PL5. Structure of the use of biological progress in grass and legume breeding by type of farmer (%) (own figure based on Polish Seed Chamber data).

On the basis of information obtained from the Polish Seed Chamber about the quantities and structure of sold seeds of grasses and legumes, it can be estimated that in the last three years that seed mixtures were sown on an area of 120,000 ha annually. Most of these mixtures were used to establish temporary grasslands. Grasslands on farms specialising in milk production are renovated systematically every five to six years. As expected, the operation is carried out more frequently if it warranted by the appearance of factors resulting in grassland degradation (Goliński, 2007). The feed base is frequently supplemented by sowing grass-legume and grass mixtures on arable land and their cultivation competes with maize. Due to the specificity of site conditions, temporary grasslands are established more frequently in northern and eastern Poland than in the central and western parts of the country where maize cultivation is preferred. The trend of increasing acreage under maize on dairy farms observed in recent years has been strengthened by the marketing of new maize cultivars, which are better adjusted to regions with shorter vegetation periods or in years with a drought period resulting in falling yields from grasslands.

There is currently a rich marketing offer of grass and legume seeds in Poland, but the mixtures of these seeds are extremely diverse even though they appear to be intended for the same purpose. Additionally, there are also frequent mistakes both with regard to the species composition and quantities of individual components to be sown. As a result, in 2004 the Polish Seed Chamber introduced standard mixtures on the Polish market intended for grasslands (Goliński *et al.*, 2003). Their application by seed companies is not compulsory but should be treated only as a recommendation.

For complementary seeding of permanent grasslands, the most suitable grass species is perennial ryegrass. Meadow fescue is used on organic soils. Options for permanent grassland renovation using oversowing or overdrilling by some dairy farmers are hybrid ryegrass, Italian ryegrass, westerwold ryegrass and festulolium. These species are applied with the understanding that they are temporary and that the permanent grassland must be systematically improved every two years. Among the legumes that are most suitable for complementary seeding are white clover on pastures, red clover on meadows located on mineral soils and alsike clover on meadows located on organic soils. In very few cases, birdsfoot trefoil and lucerne are applied. It is important to select varieties preferred for complementary seeding, which are distinguished by their ability for quick rooting and installation in old swards as well as being highly competitive.

Regarding the suitability of grasses and legumes for mixtures using in total sod renovation of permanent grasslands, the ranking of grass species (from most to least important) are perennial ryegrass, meadow fescue, tall fescue, timothy, smooth-stalked meadow grass, cocksfoot, red fescue, black bent, tall oat-grass, and others. The ranking of the legume species for mixtures (from most to least important) is white clover (particularly small-leaved varieties), alsike clover, red clover and birdsfoot trefoil.

When setting up temporary grasslands, the most productive grass and legume species are suitable. The most important grass species for pure sowing or mixtures (from most to least important) are Italian ryegrass, westerwold ryegrass, perennial ryegrass (mainly tetraploid varieties), festulolium, meadow fescue, timothy, tall fescue, cocksfoot and brome grass. The ranking of legume species (from most to least important) is alfalfa, red clover, white (large-leaved varieties), alsike clover, crimson clover, Persian clover, birdsfoot trefoil, Egyptian clover and common sainfoin.

The outcome grassland renovation following improvement of the botanical composition of the sward should be an increase in both yields and herbage quality; these improvements may occur jointly or separately. The resulting forage yield increase or quality improvement should fully offset the costs incurred for the renovation. Numerous economic analyses indicate that the improvement of herbage quality from grasslands after renovation is justified by milk production. Goliński (1998) maintains that in view of the progressing degradation of meadows and pastures, the ratio of the production costs of 1 kg DM, 1 MJ energy and 1 kg protein of feed to the upgrading value of feed in animal production is very high. A rational indicator of the time at which meadow oversowing or overdrilling should be done is when the value of the ratio of the production cost of 1 MJ feed energy to the upgrading value of feed in dairy cows production reaches 1.2 to 1.5. A later application of complementary seeding can be justified if the old sward is increasingly damaged (mechanically or chemically) and the amounts of seeds needed for oversowing or overdrilling continually increase.

However, further grassland degradation associated with an increasing ratio of the production cost of 1 MJ energy to the upgrading value of feed requires renovation involving resowing.

▸▸ References

Baryła R., 2001. Changes of the species composition of meadow undergrowth in post-boggy habitat (a synthesis of 30-year-long studies conducted in Sosnowica, the Wieprz-Krzna Canal Region). *Annales UMCS* E, 56, 65-79.

Dąbrowska-Zielińska K., Goliński P., Jørgensen M., Mølmann J., Taff G., Tomaszewska M., Golińska B., Budzyńska M., Gatkowska M., 2015. New methodologies for grasslands monitoring. *In: Sustainable use of grassland resources for forage production, biodiversity and environmental protection.* Vijay D., Srivastava M.K., Gupta C.K., Malaviya D.R., Roy M.M., Mahanta S.K., Singh J.B., Maity A., Ghosh P.K. (eds), Proceedings of the 23rd International Grassland Congress, New Delhi, 30-40.

Goliński P., 1998. Factors affecting complementary seeding of grasslands – economics. *Grassland Sciences in Poland,* 1, 65-78.

Goliński P., 2007. Grassland renovation in Poland. In: Conijn J.G (ed.) Grassland resowing and grass-arable rotations. Plant Research International B.V., Wageningen, 19-31.

Goliński P., Warda M., Kaszuba J., 2003. Fodder standard mixtures for grassland. *Plant breeding and seed production*, 4, 32-36 (in Polish).

Grzyb S., Prończuk J., 1995. The classification and valuation of meadow habitats and assessment of their production potential. In: Trends in the development of grassland science in relation to the current state of knowledge in its key areas. Materiały Ogólnopolskiej Konferencji Łąkarskiej. Wydawnictwo SGGW, Warszawa, pp. 51-63 (in Polish).

GUS, 2018. Agriculture in 2017. Statistical Yearbook. Main Statistical Office, Warszawa, pp. 202. (available at www.stat.gov.pl)

Jankowska-Huflejt H., Domański P., 2008. Present and possible directions of utilising permanent grasslands in Poland. *Water-Environment-Rural Areas*, 8 (2b), 31-49.

Kozłowski S., Stypiński P., 1997. The grassland in Poland in the past, present and future. *Grassland Science in Europe* 2, 19-29.

Kryszak A., Grynia M., 2005. Grass communities of excessively wet sites in river valleys. *Grassland Science in Poland* 8, 97-106.

Kucharski L., 1999. The flora of the meadows of central Poland and its changes in the 20th century. Wydawnictwo Uniwersytetu Łódzkiego, Łódź, pp. 168.

Nyc K., Pokładek R., 2008. The current problems of grassland drainage and reclamation. *Water-Environment-Rural Areas* 8, (2b), 97-103.

Warda M., Kozłowski S., 2012. Grassland – a Polish resource. *Grassland Science in Europe*, 17, 3-16.

Warda M., Stamirowska-Krzaczek E., 2010. Floristic diversity of chosen grass communities in the Nadwieprzański Landscape Park. *Annales UMCS* E, 65, 1, 97-102.

Wasilewski Z., 2011. The efficiency of dairy cow grazing in a large farm. *Water-Environment-Rural Areas* 11, 2 (34), 173-180.

Zastawny J., 2000. Country pastures/forage resource profiles. Poland. FAO, (available at www.fao.org/ag/AGP/ AGPC/doc/Counprof/Poland.html).

France

Julien Fradin

▸▸ Introduction

France has 27 million ha of utilised agricultural area (UAA), including 12 million ha where grain crops are grown and 11 million ha of grasslands. The country also has 20 million ha of forests (BPSCA, 2018). With regard to its geography, mainland France has various climates with a range of influences (oceanic, continental, mountainous and Mediterranean) along with different types of soils. France benefits from a highly diversified agricultural offer due to this combination of pedoclimatic conditions.

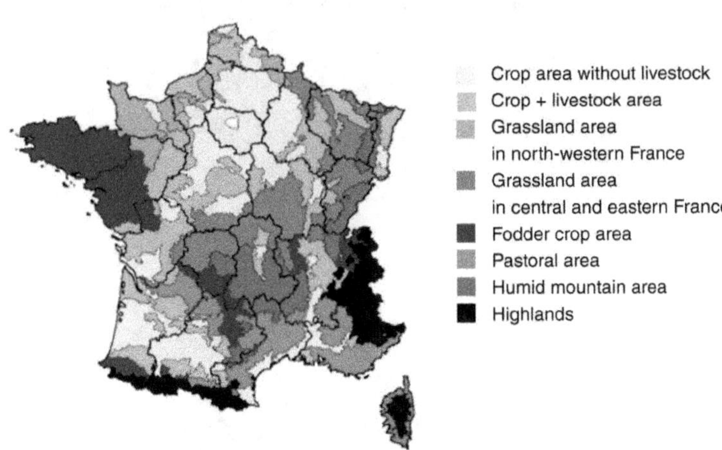

Figure FR1. The major livestock rearing areas in France.

Based on the presence or absence of livestock husbandry, the dominant crop rotation of each region, soil potential and altitude, the main regional farming systems can be divided into eight categories (see Figure FR1). There is a clear distinction between lowland areas and higher altitude areas. In the lowlands, certain areas of France are dominated by arable crops where ruminant livestock is mostly absent. Intermediate areas, which often have lower grain production potential, have developed integrated cropping and livestock systems. The intensive agricultural production areas of western France and Piedmont, where maize silage plays an important role in the farming system, are dominated by dairy farming in general, or suckling veal production in the western Massif Central. Grassland areas in the lowlands (north-western, eastern and central France) are characterised by a higher proportion of permanent grasslands where dairy and suckler cows are both raised. Finally, the pastoral zones, located at various altitudes, are defined by a very extensive stocking rate utilising semi-natural grasslands and rangelands with strong heritage value. With regard to the high altitude areas, humid mountain areas are characterised by altitudes of less than 2000 m asl and have a significant proportion of permanent grassland areas (+80% of UAA). These areas are specialised in dairy production (with the exception of the Massif Central, where suckling to weanling systems are also important), which is well-known in France for its use in cheesemaking. In high mountain areas, livestock are transhumant with pastures used for grazing in summer. These areas also often benefit from agro-tourism and high value-added products.

▸▸ Herbivore stocking rate

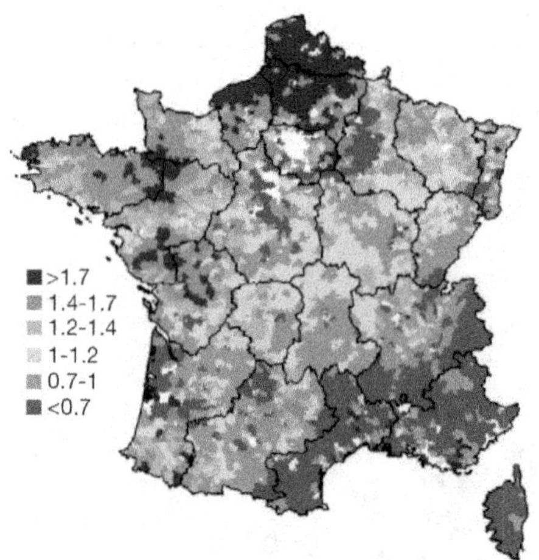

Figure FR2. Herbivore stocking rates LSU/ha FA (GEB – Institut de l'Elevage, 2013).

In 2010, the average herbivore stocking rate in France was 1.2 LSU/ha of forage area (FA). Stocking rates are highest in the lowlands and Piedmont regions where milk production is high. This phenomenon is exacerbated when there is competition on arable land with high value-added crops (potatoes, sugar beets etc.), such as in the north of France. Overall, the northern half of France benefits from favourable pedoclimatic conditions and can support above-average stocking rates. Areas where suckler herds thrive have lower stocking rates and are generally located in disadvantaged areas. In the humid mountain regions, the overall stocking rate is about 0.9 LSU/ha in relation to grass growth over a short period. Pastoral and high mountain areas have the lowest stocking rates with less than 0.7 LSU/ha FA.

▶▶ Grassland-based livestock farming systems

Dairy herds

The geographic distribution of dairy farms follows a horseshoe shape, starting in the west, where density is highest, moving through the integrated cropping and livestock areas of the north and the east and ending in the south-western part of the Massif Central.

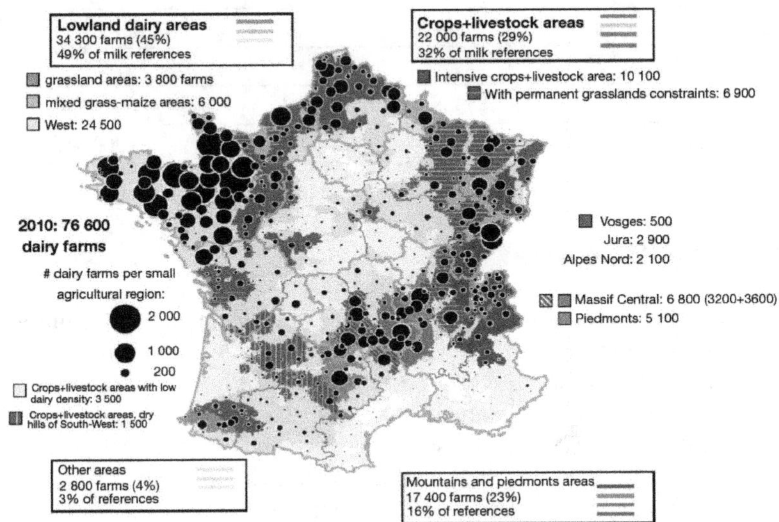

Figure FR3. Classification of French dairy system (source: Agreste, RGA 2010, analysed by the Institut de l'Elevage).

▶▶ Milk production

For 2014/2015, the milk collected from 65,600 farms totalled 25.1 billion litres, an average of approximately 383,000 l of milk per farm. Since 2007, the average amount of milk collected per farm has increased by about 18,000 l per year, while the number of farms has fallen by about 4,000 per year (FranceAgriMer, 2016). Over this same period, milk collection showed a slight increase. The seasonal variation in production is about 20% between the highest and lowest collection periods (i.e. May and September, respectively) (GEB-Institut de l'Elevage, 2018). Most milk collected comes from lowland areas (51.6%) and mixed-livestock areas (31%) compared to only 15% from mountain and Piedmont areas. The trend is reversed when it comes to collecting milk for high value-added products (PDO, PGI). In fact, 87% of PDOs come from mountain and Piedmont regions. France has 28 cheeses, two creams and three butters registered as dairy cow PDOs, along with PGIs for nine cheeses and one cream. In 2017, about 11% of the milk collected is used for PDO and PGI product, (10.3% and 0.9%, respectively).

Table FR1. Contribution of areas to dairy production and their productivity per forage area (Brocard *et al*, 2015).

Areas	% farms	% deliveries	% specialised	Milk per forage area (l/ha)	% dairy PDOs
LDAs	46.5	51.6	37	6,600	3
West	*33.2*	*37.3*	*42*	*7,000*	*1*
CLA areas	28	31	23	7,400	4
Intensive	*13.4*	*14.7*	*22*	*8,600*	*2*
MPAs	*22*	*15*	*67*	*3,700*	*87*
Jura	4	3.2	84	*3,000*	38
Other areas	3.5	2.4	41	5,300	6

▸▸ Dairy herd size

Before the introduction of milk quotas (1983), France had 7.2 million dairy cows versus 3.8 million in 2017. Over this period the decline was a slow but continuous process. The mountain and Piedmont areas have been less affected than lowlands near crop areas. This is explained by the presence of higher value-added products in these regions that allows for a less competitive and more robust situation for farmers. The decrease in the number of dairy cows did not affect the overall volume of milk collected thanks to improved individual performances.

Holstein Friesian is by far the most common breed in France with 2.4 million, followed by Montbeliard and Normande breeds with nearly one million cows. Holstein Friesians produce more milk per cow, but the other two breeds provide greater added value. Crossbreeds are still uncommon in France.

Nearly 70% of French farms have fewer than 60 dairy cows but 75% of dairy cows are owned by farms with more than 50 dairy cows. In 2018, 11% of farms had more than 100 dairy cows.

▸▸ Milk production per cow

Depending on the feeding systems, production per cow varies, with 6,000 l/cow in lowland grazing systems with one tonne of concentrate to up to 8,500 l/cow with 1.6 tonnes of concentrate and 44% maize silage in the annual ration for integrated cropping and livestock systems. Mountain and Piedmont areas have a per-cow milk yield that is slightly higher than the lowland grassland system but with a higher proportion of concentrates in the diet (225 g/l vs 166 g/l). These averages per cow hide large variations in production within these groups (Observatoire des élevages laitiers, 2015).

Accessibility of grazing areas

In France, 92% of dairy cows are grazed, with 87% grazing more than 170 days per year and 71% of farms offering more than 0.2 ha of pasture per cow. The contribution of grazing in the annual ration of dairy cows varies from 10% or 700 kg DM for systems maximising maize silage versus 31% or 1,800 kg DM for grassland systems. French dairy farms have on average 31 ha of permanent grassland and 22 ha of temporary grassland. There are, however, significant disparities within and between dairy areas. While the tendency is for herd expansion (+2 dairy cows/year/farm on average), accessibility of grazing areas does not follow the same path. Land expansion is needed for growing home forage, but it is often far from the milking parlour and the accessible grazing area per dairy cow generally decreases over the years. As a result, the larger the herd, the lower the contribution of grazing to the diet. Meanwhile, grassland systems with more than 0.8 ha accessible grazing area per cow have increased, which could be related to the increase in organic dairy farming.

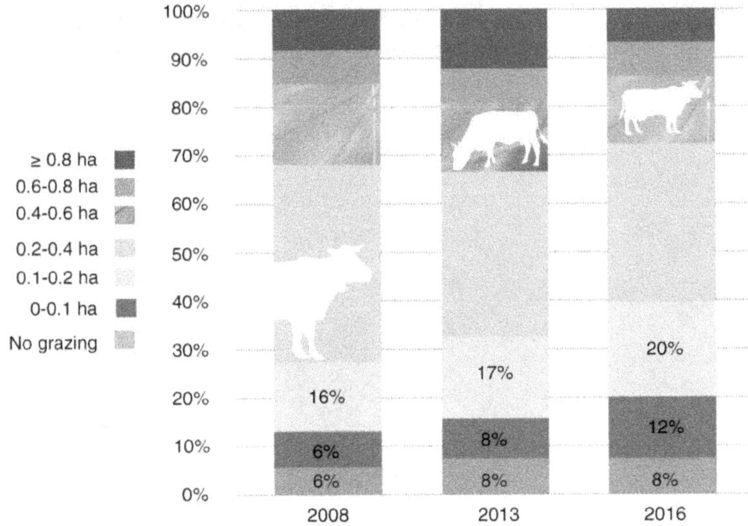

Figure FR4. Changes in accessible grazing area per dairy cow (Observatoire des élevages laitiers, 2018).

Rise in automatic milking systems

As in other dairy countries, the use of AMS in France is growing, with about 4,800 in 2015. For every new milking parlour built since 2011, half use milking robots.

Installing a robot often leads to a decrease in or even stoppage of grazing, even though many studies have shown the ability to keep grazing sustainably with a milking robot. In 2017/18, 34% of AMS were fully housed.

Organic dairy farming

Following past declines in milk prices, there has been an increase in conversion to organic dairy farming. In 2016, 2,500 farms were organic, mostly located in western France and in humid mountain areas where the grassland growing period is longer during the season. With an average of 48 cows per farm, the milk volume is estimated to be 230,000 litres per farm, with the forecast for 2019 at nearly one billion litres of organic milk. This would mean an increase in milk collected of 75% between 2016 and 2019. To date, demand for organic products in France has always absorbed the increasing production.

Suckler herd

The suckler herd in France is associated with the two main breeds: Charolais and Limousin. Suckler cows are located in the traditional Charolais and Limousin breeding area, which comprises the grassland area of the northern Massif Central with 1.01 million suckler cows. The intensive agricultural area of western France (530,000 cows) also has a high density of suckler cows, as does the Piedmont area of the south-western Massif Central and the Pyrenees. Suckler cows are spread throughout grassland and forage areas.

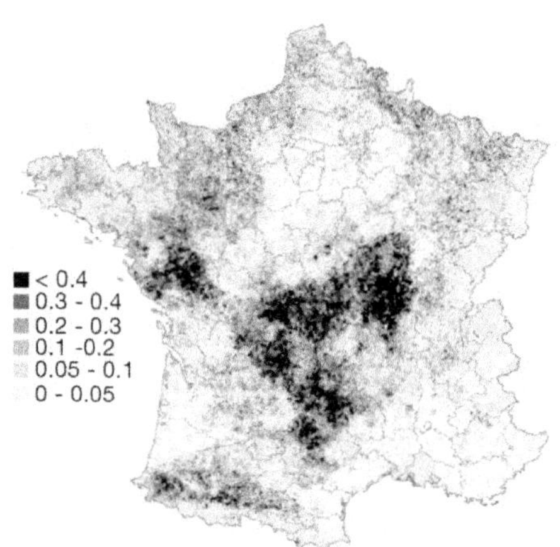

Figure FR5. Number of suckler cows per ha in townland (GEB – Institut de l'Elevage, 2013).

▸▸ Herd size and variation

Before the milk quotas were establishmed, France has 2.9 million suckler cows. Growth was strong until the 2000s, when it levelled off to the now-stable national herd number of about 4 million suckler cows. All regions have been affected by this phenomenon. Half of this growth was in the dairy zones and half in the traditional areas. In 2018, Charolais was the most common breed with 1.45 million dams, followed by Limousin with 1.1 million cows, Blonde d'Aquitaine with 0.48, and finally the more rustic Salers and Aubrac breeds, which together account for 0.4 million cows.

In 2017, there were about 85,000 farmers with suckler cows; this figure has dropped from 105,000 just 10 years ago. Among these farmers, 44% have fewer than 29 cows. However, 1.9 million suckler cows are held by 21% of farmers with herds larger than 70 suckler cows.

▸▸ Characteristics of production

Meat production systems in France are mainly breeders and not finishers. This reflects the fact that the vast majority of farmers hold herds of breeding cows, whose main purpose is to produce late-weanlings and cull cows. The most common system, found on about two thirds of arms, produces lean animals to be fattened in specialised systems. Such fattening systems do exist in France, but most are located in neighbouring countries, namely Italy and Spain. These breeders sell weanlings for the store market and cull cows for butchery. In the Charolais area, some weanlings are kept longer and fed with cut grass and cereals to sell heavier weights for the store market. Traditionally in this region, a lean young bull is grazed until the end of its second spring before being sold to a finisher.

About 20% producers, mainly in lowland area, have developed suckling to beef systems. By creating a fattening system, breeders increase the amount of meat sold per LSU on the farm. These workshops go hand in hand with the introduction of maize silage used for fattening bulls born from the suckler cow herd. They have come to replace the more traditional suckling-to-steer beef systems because they are faster to finish.

Steers are no longer common in France. They rely on grassland systems. Maize is almost never used, especially for the image of the final product. These systems are not very demanding in terms of labour because pasture is predominant in the diet. They are usually found on large farms with land constraints.

Finally is the system for producing suckling veal. This system is mainly found in south-western France, the traditional region for this product. Calves are housed while cows are grazed or fed separately, with suckling twice a day. The practice lasts three to 5.5 months to obtain a carcass weight of 85 to 170 kg. Margins are the highest of all beef production systems with equivalent stocking rates, but the system is very time consuming. This system of production is aimed at producers who, having a forage surface constraints, cannot move into weanling production.

Sheep

Dairy

Sheep production is regionalised and consistent with the local environment and animal production. Sheep farms specialising in dairy production are mainly located in Roquefort area, with a high added-value PDO and in the western Pyrenees, namely the Ossau-Irraty PDO. There are also flocks in south-eastern France which create value from cheese-making on the farm.

In 2010, 5,000 farms with dairy sheep had a total of 1.4 million ewes. These farms produced 270 million litres of milk, with two-thirds collected within the Roquefort area. The amount of milk collected was rising for several years before levelling out over the past 10 years.

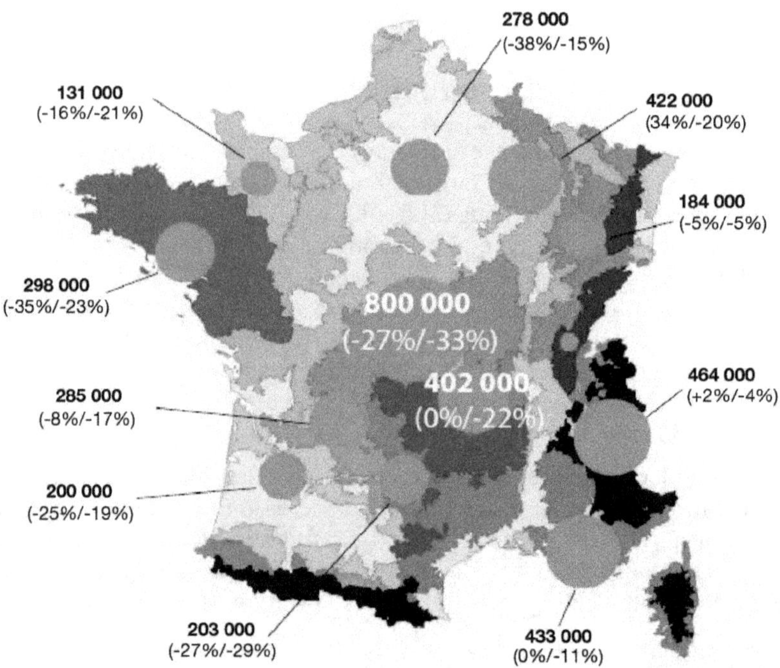

Figure FR6. Number of ewes for meat in 2010 (4 142 872, -21%/-20%) and their evolution 1988-2000/2000-2010 (RGA 1988, 2000 and 2010, analysed by the Institut de l'Elevage).

In the Roquefort area, farms have an average of 81 ha of UAA, 80% of which is dedicated to forage production. Permanent grassland areas account for 19% of the forage area, while temporary grasslands, mainly grass-legume with high lucerne content, accounts for 79% of the FA. The non-forage area is dedicated to growing cereals for energy concentrate and straw for the flock. The flock is growing steadily on these farms and the average stocking rate is 1.3 LSU/ha, which comes out to about 300 ewes per farm on average. In this region, gross margins are highly dependent on milk production per ewe, with an average of 261 litres per ewe.

In the western Pyrenees, farms are smaller, with an average of 33 ha. Forage area represents 98% of the UAA. Permanent grasslands account for 59% of the FA and temporary grasslands account for 36%. Maize is grown on every other farm and occupies less than 10% of the FA. The average flock is 63 LSU (about 225 ewes) for an average stocking rate of 2 LSU/ha. These high stocking rates are consistent with the local pedoclimatic conditions, which make it possible to ensure strong forage production that is well distributed throughout the year.

Meat

Farms that produce sheep meat are concentrated in the southern part of the country. In contrast to dairy production, lamb production has suffered since 1980 with a sharp drop in the number of both farmers and ewes. This is partly explained by the specialisation of farms, which gave up mixed farming to concentrate on either cropping, milking or increasing suckler cow herds.

Many farms have small flocks. For example, 66% of all ewes are held by 16% of farms with flocks of more than 200 ewes. In 2017, only 3.6 million ewes remain in the national flock, a decline from 6.5 million in 1988. To be able to address future needs, the meat sheep sector faces a number of challenges, including the post-Brexit context, lower consumption, increased supply of the domestic market (45% today), difficulty attracting young farmers, maintaining natural areas for habitat conservation through grazing. When looking at the numbers of sheep farms, two major characteristics are essential for economics. The first is the importance of lamb mortality, and there are very large differences between farms. In 2009, the median value was 16%. The second characteristic is the strong relationship between the margin of feed cost per ewe and productivity, expressed in kilograms of lamb per ewe.

Between regions, stocking rates vary from 1.1 LSU/ha FA in grassland areas in the Charolais area to 2.5 LSU/ha FA in integrated cropping and livestock areas, thanks to higher content in the diet. The latter system may also use cover crops to feed ewes at times when there is no commercial crop being grown. Hay is the main form of forage preservation, except in mixed systems where silage is available. There is a strong contrast between grassland specialists with very little grass stock, sheep production relying heavily on grazing, and those in high mountain areas with large stocks.

▶▶ The importance of grassland in France

Over a period of five decades, there was a clear decline in grasslands and forage crops, with near total disappearance of fodder crops (fodder beet), a very pronounced decline of forage legumes as pure crops, a significant decline in permanent grasslands (especially areas still in productive grasslands that are neglected), the rise of silage maize, and a slight increase in temporary grasslands.

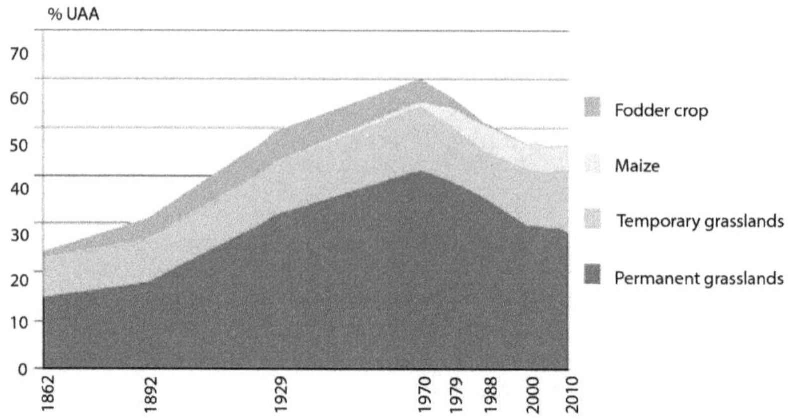

Figure FR7. Evolution of the forage area in France (GEB – Institut de l'Elevage, 2013)

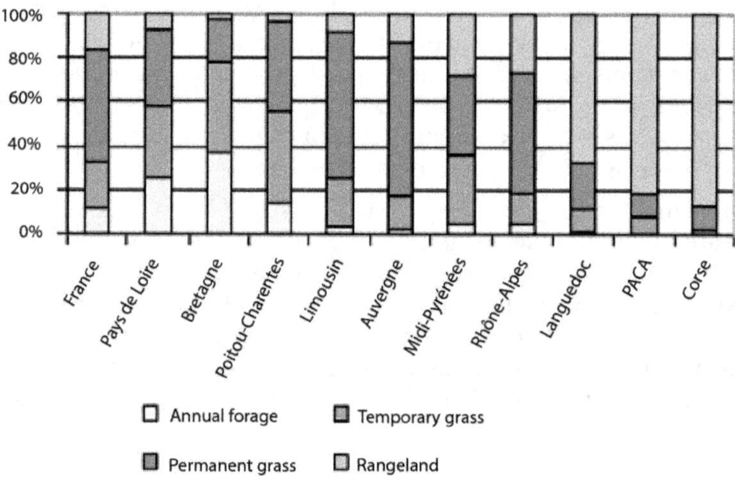

Figure FR8. Composition of forage area in French regions (Huyghe *et al.*, 2015).

The temporary grassland composition deeply changed with a very large proportion of current areas in mixtures and associations, whereas monocultures predominated in 1960.

The French territory appears very heterogeneous with regard to the different types of grasslands that are cultivated. In the western and central-western regions, annual forages and temporary grasslands predominate. In mountain regions, it is mostly productive permanent grasslands that support the many herds of cattle. Finally, in the south-east, the rangelands (low productive permanent grasslands) constitute the bulk of the grasslands.

Temporary grasslands

In 2001, grass-legume mixtures accounted for about 45% of the area sown on temporary grasslands. Perennial ryegrass and white clover mixtures alone accounted for 28% of the temporary grassland area. White clover is nearly always sown on temporary grasslands, especially for grazing. Multi-species swards with a minimum of two grass species and two legumes accounted for 18% of these grasslands with a very large regional disparity: they cover 3% of temporary pastures in Brittany and 56% in Limousin.

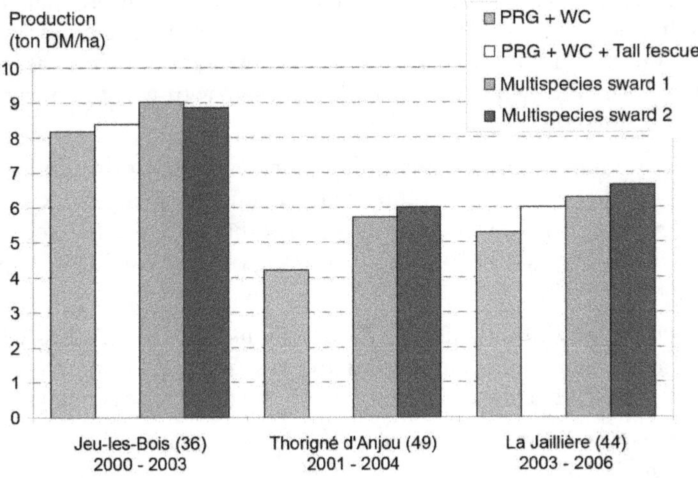

Figure FR9. Yield comparison of multi-species swards versus a grass-legume mixture (Protin *et al.*, 2014).

Over the last 20 years, the proportion of temporary grasslands sown with mixtures has steadily increased, exceeding 75% of sown grasslands today. They are predominantly sown with grass-legume mixtures. Perennial ryegrass and white clover mixtures dominate in the north-west (Brittany, Normandy). In the other regions, the situation is more diverse, especially when the proportion of hay and silage increases. In such cases, Italian ryegrass, cocksfoot and tall fescue may contribute a significant share of the grasses. Lucerne is also increasingly being sown in mixtures along with cocksfoot and tall fescue and then harvested as hay. Such a mixture is of high feeding value for sheep and goats.

Studies comparing multi-species swards to mixtures of perennial ryegrass and white clover have shown yield advantage for the multi-species swards of about one tonne of DM per ha. This advantage tends to be even greater during drought periods. From a nutritive point of view, they have slightly lower overall energy (-0.05 UFL) and crude protein values (-1% CP). This effect is greater during heading because grass species head at different time, therefore making the optimum stage harder to obtain for the whole sward.

Permanent grasslands

Permanent grasslands are an important part of forage areas in France. However, although it is known that they are favourable for biodiversity and that are generally high in species richness, their actual botanical composition is not well understood. In 2009 and 2010, Launay *et al.* (2011) studied a network of 190 permanent grasslands. These permanent grasslands were used in different livestock systems (beef, lamb, dairy cattle) and on a wide range of pedoclimatic conditions, from the Atlantic coast to highland meadows, up to 1,200 m asl. Annual production, based on late spring, summer and autumn cuts, was estimated at 6.2 t/ha/year, with very large variations: 25% of meadows produced less than 4.2 t/ha/year and 25% produced more than 8.1 t/ha/year. Spring production provided 75% of average annual production on these permanent grasslands. Summer and autumn productions show a greater contrast between regions. In fact, over the two years of the study, regrowth was too weak to justify a harvest in more than 25% of the grasslands.

For all of the permanent grasslands studied, organic matter digestibility averaged 77% at the beginning of spring and 64% at the end of spring, which corresponds to the pasture stage for the first and the end of heading for the second. The average protein content was 17% at the beginning of spring, while it was 9% at the end of spring.

The combined analysis of dry matter production, its distribution throughout the cycles and the feed value confirmed the negative relationship between production and nutritive value, which declines more rapidly on the most productive grasslands.

As a result, five main characteristics differentiate feed value on permanent grasslands (Baumont *et al.*, 2012):
– The ability to have strong spring production and total production for the year. These permanent grasslands can provide forage stocks and/or be an excellent source for spring grazing.
– The ability to grow in the summer and fall, which is partially independent of the spring production potential.
– The ability to produce high-quality forage for early harvest and regrowth. These grasslands, which are rich in legumes, then become a source of quality forage, provided they are grazed at fairly early stages.
– The ability to produce quality forage in late spring. These grasslands, rich in legumes and broadleaf, have few grasses and low productivity.
– Stability of quality throughout the spring. This has been encountered in the case of grasslands with low dry matter production with a significant presence of non-leguminous dicotyledonous. These grasslands can then be utilised for long periods without a large variation in quality.

Grazing strategy

In France, there is a wide range of terminology to describe the types of grazing practised. The most frequently use are set-stocking and rotational grazing.

Key figures to keep in mind for set-stocking are:
– Grass height (compressed) between 6 cm and 8 cm
– Between 0.3 and 0.8 ha of available area per animal

Pros	Cons
Little infrastructure needed (fence, pathway, water trough) Simplification of labour High individual performance	Higher land requirement per animal (+20%) Less production per ha Sensitivity to dry spell

Rotational grazing is based on the principle of a relatively short grazing time per paddock and a long pasture resting time depending on the season to offer a greater quantity of and higher quality grass. There must be multiple fenced paddocks.

Key figures to keep in mind for rotational grazing are:
– Rest grass from 18 to 21 days in spring and 28 to 35 days in summer when growth slows
– Entrance height: between 8 cm and 15 cm
– Residual height: between 3 cm and 6 cm (depending on entrance height)

Pros	Cons
Increase pasture utilisation More stock harvested on the same area Respects the grass growth curve	High infrastructure cost (fence, pathway, water trough) High skill level needed for paddock management Production variation during grazing (except if less than one day)

Good grazing management is possible regardless of grazing strategy, but good observation and adaptation to grass growth is necessary.

▸▸ References

Brocard *et al.*, 2015. High output farming systems in Europe: the French case.

Bureau des Statistiques sur les Productions et les Comptabilités Agricoles, SAA 2016 – 2017 provisoire, http://agreste.agriculture.gouv.fr/IMG/pdf/saa2018T1bspca.pdf

Chambre régionale d'agriculture des Pays de la Loire, 2007. La prairie multi-espèces, édition mai 2007.

Delaby L., Huyghe C., 2013. Prairies et systèmes fourragers, Editions France Agricole.

FranceAgriMer, 2016. Évolution des structures de production laitière en France, édition mars 2016.

GEB – Institut de l'Élevage, 2013. Dossier Économie de l'Élevage n° 440-441 - Novembre-Décembre 2013.

GEB – Institut de l'Élevage, 2018. Les chiffres clés du GEB, bovins 2018 production lait et viande.

Huyghe C., Peeters A., de Vliegher A., 2015. La prairie en France et en Europe.

Launay F., Baumont R., Plantureux S., Farrie J.-P., Michaud A., Pottier E., 2011. Prairies permanentes – Des références pour valoriser leur diversité.

Leray O., Doligez P., Jost J., Pottier E., Delaby L., 2017. Présentation des différentes techniques de pâturage selon les espèces herbivores utilisatrices, journées AFPF, mars 2017.

Observatoire de l'alimentation des vaches laitières, 2018. Références, Edition 2015 – 2018.

Observatoire des élevages laitiers, 2018. Le pâturage des vaches laitières françaises, édition 2018.

Protin P.V., Pelletier P., Gastal F., Surault F., Julier B., Pierre P., Straëbler M., 2014. Les prairies multi-espèces, une innovation pour des systèmes fourragers performants, journées AFPF, mars 2014.

Italy

Claudio Porqueddu, Rita A.M. Melis,
Lorenzo Pascarella and Riccardo Negrini

▶▶ Introduction

Italy is a peninsula extending far into the Mediterranean Sea. The estimated population is 60.4 million people (data 2017). The population is unevenly distributed. The Po Valley, in northern Italy, and the metropolitan areas of Rome and Naples, in central and southern Italy, are densely populated, while the Alps and Apennines highlands, as well as the island of Sardinia are very sparsely populated. The average population density is 201 inhabitants per square kilometre, a value that is higher than that of most western European countries.

Italy has a surface area of 301,338 km^2. The territory is mostly hilly (42%) with plain areas only covering 23%, mainly in the Po Valley, which is the country's largest plain (44,000 km^2). The rest of the territory is covered by mountain areas (35%). The Alps, along the northern border, and the Apennines, which form the backbone of the entire peninsula and the island of Sicily, are the largest mountainous systems. Italy lies in the temperate zone but there is a strong climate variation between the northern regions bordering the European continent, and the south, surrounded by the Mediterranean Sea, due to the considerable length of the peninsula (1,200 km). Three main biogeographic regions dominate the country:
− the Alpine area, which has a continental mountain climate;
− the continental climate area, mainly represented by the Po Valley, characterised by hot summers and severe winters;
− the Mediterranean region, covering the central and southern Italian regions and the islands of Sicily and Sardinia. The most important climate trait is the concentration of rainfall during the relatively mild winters and its total absence during the hot summers, associated with a large intra- and inter-annual variability.

Italy has a variety of soils as well, originating from several types of parent materials, at different altitudes, under different climate conditions in preceding eras. Dark-brown podzols are very common in the Alps, where rainfall is heavy. In the Apennines, brown podzolic soils predominate, supporting forests, meadows and pastures. Rendzina soils are characteristic of the limestone and magnesium limestone mountain pastures and of many meadows and beech forests of the Apennines. Sparse red-earth soils, rocky soils, clays, dune sands and gravel can be also found.

▶▶ The agricultural sector and animal husbandry with a focus on ruminants

(Modified from Lombardi *et al.*, 2012 and Porqueddu *et al.*, 2017)

Italy enjoys an abundance of agricultural resources and its potential productivity is high. It is a world leader in olive oil production and a major exporter of rice, tomatoes, wine, fruit and vegetables. Despite this, its agri-food trade balance is negative, and Italy is a net importer of agricultural raw materials but a net exporter of processed foods (Centro Studi Confagricoltura data, 2018). With regard to the zootechnical sector, meat production is not sufficient for internal demand and meat is imported from other European countries, particularly from Ireland and Germany. Italy is also quite weak in the dairy farming sector and its milk deliveries contribute 8.2% to total European milk deliveries. Italian self-sufficiency in milk is 84.5% (CLAL data, 2017). The total available milk is mainly processed into PDO cheese (Figure IT1) of which Italy is the main European producer. The exports of dairy products concern 29.3% of the national delivered milk, expressed in milk equivalents, equal to 4.7% of the European exports.

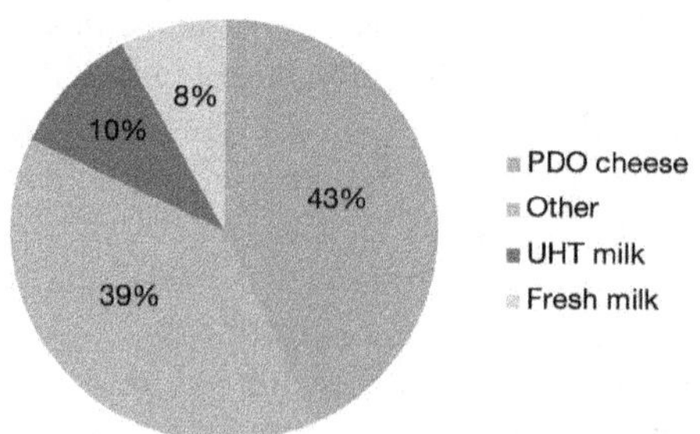

Figure IT1. Destination of total milk available in Italy. National deliveries* + Imported Bulk milk = 13 820 345 tonnes.

* Deliveries buffalo milk, goats milk and sheep milk. % fresh pasteurised milk on National deliveries: 9%. Source: CLAL data, 2017.

Table IT1. Evolution of animal heads per category and milk production in Italy. Eurostat data (2017) and CLAL data (2017).

Animal heads

Category	Year								2017/2010 (%)
	2010	2011	2012	2013	2014	2015	2016	2017	
Live bovine animals (no. × 1,000)	5,833	6,252	6,252	6,249	6,125	6,156	6,315	6,350	+8.87
Dairy cows	1,746	1,755	2,009	2,075	2,069	2,057	2,060	2,040	+16.84
Buffalos	365	354	349	403	369	374	385	401	+9.77
Beef cattle				2,084	2,014	2,042	2,085	2,123	*+1.87
Live sheep	7,900	7,943	7,016	7,182	7,166	7,149	7,285	7,215	-8.66
Milk ewes	5,416	5,469	5,302	5,247	5,142	5,137	5,206	5,130	-5.29
Live goats	983	960	892	976	937	962	1,026	992	+0.93

Milk production

Category	2010	2011	2012	2013	2014	2015	2016	2017	
Raw cow milk from farms (tonnes x 1,000)	11,399	11,298	10,876	10,701	11,037	11,161	11,524	11,950	+7.01
Sheep milk (t)			406,177	383,836	372,526	397,500	424,840	427,430	+5.23
Goat milk (t)			27,940	27,489	28,463	33,200	31,730	37,050	+32.61

	2010	2011	2012	2013	2014	2015	2016	2017	
Cow milk yield (kg/year)							6.326		
Sheep milk yield (kg/year)							172**		
Goat milk yield (kg/year)							135***		

**Data refer to lactating ewes of the Sarda breed; ***Data refer to goats bred in Sardinia.

The ruminant population consists of 6 million head of cattle, 7 million head of sheep and 1 million head of goats, which represent 7%, 8% and 8% of the total European heads, respectively. Dairy cows prevail among cattle (Italian Holstein and Brown Swiss breeds), followed by dual-purpose milk-beef breeds (e.g. Italian Simmental, Modicana, Tyrolean Grey), and, finally, specialised beef cattle (usually indigenous cattle breeds, e.g. Piemontese, Chianina, Marchigiana, Romagnola). The most important sheep breeds are Sarda and Comisana, specialised in milk production. The number of sheep showed a negative trend over the last seven years (-8.66%), while the number of cattle (+8.9%) and goats (+0.93%) both increased.

Currently, 4.2 million live head of bovine animals and 1.2 million dairy cows (65% and 60% of the national figures, respectively) are raised in the northern regions with intensive systems, where 40% of livestock holdings can be found. Meanwhile, 78% of buffaloes, 72% of sheep and 64% of goats are bred in southern Italy and the main islands (Sardinia accounts for 45% of sheep and 22% of goats).

The number of livestock holdings is about 155,000 (Eurostat, 2016). Their number dropped drastically from 2010 (-65%), and the decrease mainly concerned holdings located in mountain areas of central and southern Italy and small-sized farms. Larger sized holdings (>500 livestock units) showed an increase in number (+9%), indicating a deep restructuring of the livestock system in Italy. The average farm size is 9 ha, but large farms can be found in mountain areas (13 ha), and smaller farms in hilly and lowland areas (8 and 6 ha, respectively). Significant differences between northern and southern regions are also seen in the number of heads per farm: an average of 54 (northern) and 28 (southern) head of cattle are raised per farm. The number of dairy cattle per farm in northern Italy is double that in southern regions (33 vs. 17), while there is one-third the number of small ruminants in northern regions (31 and 14 for sheep and goats, respectively) compared to southern regions (106 and 39).

With regard to the altitudinal distribution of cattle livestock holdings, 74% of them are located in marginal areas (34% in mountain and 40% in hilly areas). However, these holdings account for less than half of the national livestock figures, and there are major differences between northern and central-southern regions that are most likely due to the farming system in place. In fact, in northern Italy, 44% of livestock holdings are concentrated in the Po Valley, where an average of 130 head per farm are raised. The number of heads per farm decreases with altitude: farms in hilly areas raise an average of 35 head of cattle per farm, while the average for farms in mountain areas is just 19. Over the last fifty years, the structure of agricultural holdings has been highly affected by socio-economic changes. A generalised reduction of agricultural surfaces has been observed all over the country, but while the agricultural area declined by 30%, 50% of the surface of permanent grasslands and meadows was lost.

Over the same period, the number of agricultural holdings decreased by 66%. Dairy farms were especially affected by drastic structural changes, with their numbers declining by 80%, while the number of dairy cows declined by only 35%. The heads per farm increased from 8 to 22. The number of sheep and goat farms dropped by 47% and 71%, respectively, while the number of heads of both the species was almost stable. Such changes resulted in an increase in heads per farm.

▸▸ Grassland types

The utilised agricultural area in Italy is equal to 11.29 million ha, about 40% of the national area (Eurostat, 2017). Grasslands cover about 6.5 million ha.

Table IT2. Evolution of grassland areas in Italy.

Grassland type (ha)	2005	2007	2010	2013
Fodder crops – temporary grasslands	955,380	946,570	1,082,490	1,024,950
Fodder crops – Green maize	215,320	220,380	233,730	274,830
Other green fodder	832,870	849,620	835,360	914,460
Other fodder crops – leguminous plants			103,560	105,510
Fallow land	473,420	494,220	547,720	365,310
Permanent grasslands and meadows	3,346,950	3,451,760	3,434,070	3,316,430

Most grasslands are permanent meadows and pastures (Table IT2), which cover 3.3 million ha (more than 50% of the total grassland area). The southern regions and main islands account for about half of national grassland areas, represented by permanent pastures. In the northern regions, permanent pastures account for 60% of the grassland area, but permanent meadows are widespread due to more favourable soils, land morphology and precipitation distribution.

The majority of permanent meadows and pastures is located in mountain areas (60%), reaching 90% in the Alps and hilly areas (33%). The Apennine Mountains show a lower proportion of permanent meadows and pastures (50%), without differences among northern, central and southern regions. In the lowlands, permanent grasslands are scarcely present and temporary grasslands, represented by annual fodder crops (maize-silage, barley, oat, sorghum, Italian ryegrass, vetch, berseem and crimson clover) and temporary grassland (lucerne, sulla, sainfoin, white clover, red clover, grasses and their mixtures) prevail.

Permanent grasslands underwent a gradual reduction from '90s to 2000, when more than 340,000 ha were dismissed, while from 2000 to 2013 their area remained stable (EUROSTAT, 2018). The largest decrease involved pastures and meadows. Currently, permanent grasslands and pastures are widely diffused in the main islands (Sardinia and Sicily, 40.7% of their UAA), followed by the regions of north-western (32.5%), north-eastern (24%), southern (20.6%) and central Italy (18.5%) (ISTAT, 2016).

▸▸ Grass yield and grass quality

Due to the several bioclimatic areas and soil variability, Italian grasslands show a great variability in terms of production and quality as well as farming systems, where animal breeds, forage resources and level of intensification are adapted to specific environmental conditions. Despite several constraints (i.e. climate, morphology, small size of farms and flocks, lack of technical assistance), Italian grassland production is rather important, showing interesting and sometimes original models of adaptation to the specific environmental conditions, which are certainly among the most difficult in Europe.

Average annual dry matter yields for grasslands are generally quite low in the rainfed Mediterranean zones, ranging between 1 and 3 t/ha, but can reach up to 10-15 t/ha in the lowlands of northern Italy under continental climate conditions. This high variability in terms of forage production can be divided as follows: marginal lands (nearly 35% of total acreage); intensive agricultural areas (nearly 15%); and intermediate areas between extensive and intensive (about 50%).

Normally in marginal areas there are no temporary grasslands, or if there are, they cover only small surfaces. Grazing is largely dominant, and permanent pastures are the main forage resources outside of woods. Generally in these areas it is possible to find two different systems. In the uplands, utilisation by grazing is limited to short periods during summer with dairy cows in the Alps and with beef cattle, sheep and goats in the upper part of Apennines and the islands. In the more typical Mediterranean areas, there are very long periods of grazing with low stocking rates. The agro-pastoral systems, widespread in interior hilly areas with little mechanisation, are based on diversified resources; semi-natural grasslands and improved pastures coexist in the better areas. Traditionally, wooded grasslands, with up to 10-40% tree and shrub cover, are also used to support livestock production in these regions. These silvopastoral systems are based on woody pastures, but grazing-animal breeding is often associated with other agricultural activities to improve farmers' incomes (e.g. cork production in Sardinia).

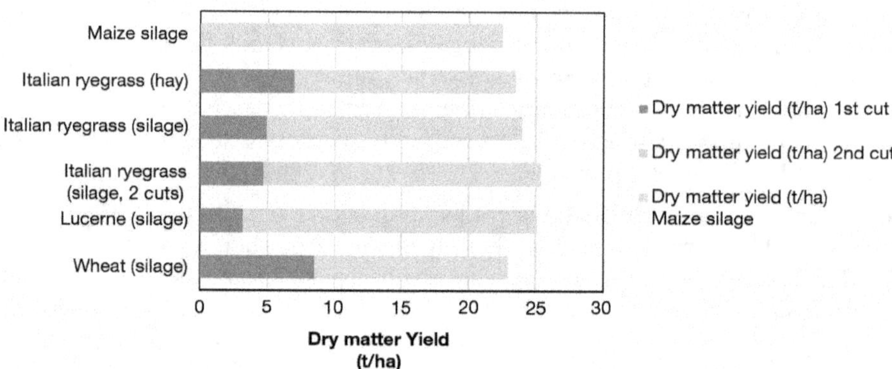

Figure IT2. Double cropping for the intensification of maize for silage systems (Borreani and Tabacco, 2016).

The intensive agricultural areas are where the greater numbers of and most productive animal husbandry systems can be found (e.g. dairy cows in the Po Valley and in other small scattered areas along littorals and Apennine valleys). The common characteristic of these areas is the zero-grazing system with the use of high amounts of supplements. The intensive forage production in the Po Valley is based mainly on maize for silage, which can also enter the forage production systems as a summer crop (double cropping) following winter cereals (e.g. barley and wheat) or Italian ryegrass as temporary grassland (Figures IT2 and IT3). Maize for silage is supported by rotated meadows of lucerne, Italian ryegrass and ladino clover or cocksfoot, tall fescue with lucerne and ladino clover, or to a lesser extent by permanent meadows. The increase of surface area sown with sorghum confirms the important role that this crop plays in the diversification of farm production. Meadows are widespread in areas where maize cannot be grown due to pedological limitations.

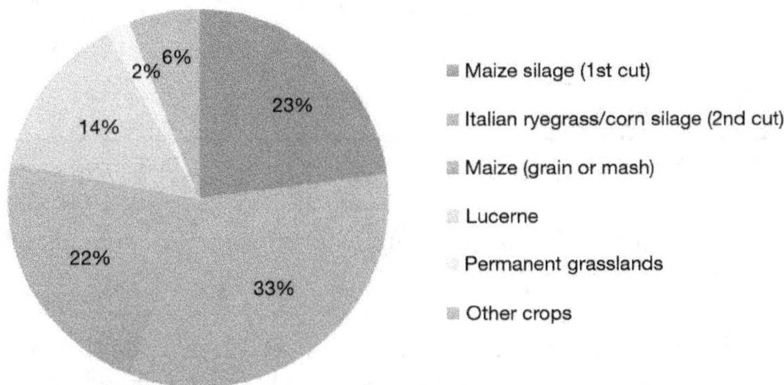

Figure IT3. Recurring forage systems in the Po Valley for dairy husbandry (Borreani and Tabacco, 2016).

The intermediate areas, between marginal and intensive zones, are where temporary and permanent grasslands coexist. They are the most spread out in the low mountains, hilly lands and rainfed plains with mainly beef cattle or dairy sheep. The forage systems are quite variable depending on the farming systems. In the coastal plains and in the dry low hills where mixed crop-livestock systems are present, feeding systems are based on a combination of annual forages and cereal stubble. In hills that can be mechanised, feeding systems based on permanent grasslands and the use of hay storage and pastures are diffused. Production systems vary from those excluding grazing up to the sharp separation of grazed areas and mowed areas, and systems based on use of grazing of the regrowth of grass-legume leys (*prato-pascoli*). The seasonal fluctuation of grassland production requires hay and/or silage reserves in order to deal with difficult periods for feeding animals.

Grassland performances in Mediterranean environments are negatively impacted by several physical constraints, which complicate the mechanisation of soil tillage, and by climate characteristics. In fact, summer drought coupled with high solar radiation levels, cool winter temperatures during the growing season, and highly erratic and variable rainfall all limit grassland productivity. As an adaptation to summer drought, annual species prevail in semi-natural Mediterranean grasslands. Their growing season ranges from four to 10 months, depending on rainfall amounts and timing and plant tolerance to water deficit (300 mm to 1,000 mm), and is characterised by two growing peaks in spring and autumn. Dry matter accumulation ranges between 110 kg/ha/day in the most favourable season (spring) and 20 to 40 kg/ha/day in autumn (Figure IT4, Caredda *et al.*, 1992).

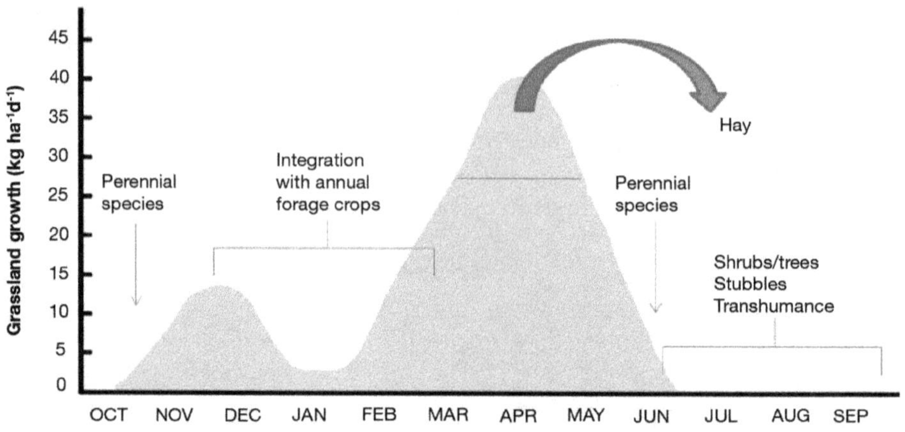

Figure IT4. Herbage production and integrations to grazing animal diets in Mediterranean areas.

Annual and inter-annual forage productions under rainfed conditions are usually extremely variable but generally limited, and depend on grassland management and soil fertility. Typically, average dry matter yields range from 0.5 to 1.0 t/ha/year in semi-natural grasslands, which prevail in marginal soils, to 6.0 to 7.0 t/ha/year, in agriculturally improved grasslands (Table IT3, Huyghe *et al.*, 2014). In grasslands subjected to shrub encroachment, herbage production declines with the increasing of shrub cover, as well as its nutritional value (Zarovali *et al.*, 2007). In the latter case, an appropriate agronomic or grazing management aimed at controlling shrubs should be introduced to promote grassland renovation and conservation. In semi-natural grasslands, forage is usually low quality, often worsened by a relatively high rate of unpalatable species. A better forage quality can be attained by applying P-fertilisers once a year to boost annual pasture legume production, but when their natural seed bank is not sufficient, the re-sowing with annual self-reseeding pasture legumes is appropriate (Porqueddu and Gonzales, 2006). The most frequently used mixtures include three to four species and are based on subterranean clovers (*Trifolium subterraneum* L. *sensu lato*) and annual medics (*Medicago* species).

To complement the insufficient pasture production in Mediterranean regions, annual temporary grasslands are widely exploited because of their high winter growth rates and flexible use. Traditionally, mixtures of annual forage legumes and winter cereals (oat, barley and triticale) or grasses (especially Italian ryegrass, *Lolium multiflorum* Lam. ssp. *italicum* and ssp. *Westervoldicum*) are used for short-term forage crops on arable lands. The most-used legume species are common vetch (*Vicia sativa* L.), woolly pod vetch (*V. villosa* ssp. *dasycarpa* (Ten). Cav.), Persian clover (*T. resupinatum* L.), crimson clover (*T. incarnatum* L.) and berseem clover (*T. alexandrinum* L.). These temporary grasslands are exclusively cut for hay production or mowed after the winter grazing (one or more grazings per season). Often, farmers harvest forage with a delay which has negative consequences on quality. Recently, farmers have introduced some mixtures based on annual self-reseeding pasture legumes and winter cereals to extend the duration of temporary grasslands to two or three years. Among perennials, lucerne represents the primary temporary grassland species for neutral and alkaline soils. Quite often the seed of local ecotypes is used in pure stands as green forage, hay or dehydrated forage (three to four cuts between June and October). Lucerne stands typically persist for three to four years under rainfed conditions or occasional irrigation before a rotational crop is grown. Despite their wide-spread natural distribution in hilly areas, perennial legumes such as red clover (*Trifolium pratense* L.) and birdsfoot trefoil (*Lotus corniculatus* L.), which are adapted to moderately acidic soils, have not been widely used. The same is true for sulla (*Sulla coronaria* (L.) Medik.) and sainfoin (*Onobrychis* spp.), although there has been renewed interest in each of these perennial legumes (Porqueddu *et al.*, 2016). A few varieties of perennial grasses, particularly cocksfoot (*Dactylis glomerata* L.), tall fescue (*Festuca arundinacea* Schreb.) and bulbous canary grass (*Phalaris aquatica* L.), are sown in higher rainfall areas with deeper soils and are generally included in seed mixtures with annual or perennial legumes.

Table IT3. Grassland dry matter yield (DMY) (t/ha) in six Sardinian sites (average of five years). Fertilisation: 100 kg P_2O_5/ha, 50 + 50 kg N/ha (from Bullitta and Caredda, 1982).

Site	Altitude (m asl)	Type of soil	DMY (t/ha)		DM %	Extension of forage availability (in weeks)
			Unfertilised	Fertilised		
Bonassai	80	Limestone	4.23	8.23	95	+7
Chilivani	350	Alluvial	2.77	5.05	82	+7
Badde Orca	600	Trachitic	3.13	5.52	76	+3
Pattada	650	Granite	4.44	6.33	43	+4
Campeda	650	Basaltic	3.92	6.41	63	+3
S. Antonio	650	Basaltic	2.39	5.38	122	+8

▸▸ Grazing

In northern Italy, generally, grazing is mainly carried out for a limited period of the year, from two to four months. In central Italy, this phase can easily last up to six months, while in southern Italy and the islands, the mild winters allow year-round open-air grazing. Both continuous and rotational grazing are used, and rationed grazing is also used with limits to grassland access (two to four hours per day). In Table IT4, the suggested values of grass height for entry and exit from the grazing sector in some grasslands used by sheep with rotational grazing are reported. The grazing system can also change during the year in relation to the farm structure, farm grassland resources, grass seasonal growth and animal physiological stage (Table IT5). Hereafter, some of the most common grazing-based farming systems in the country are described.

Table IT4. Suggested values of grass height for entry and exit from the grazing sector in certain grasslands used by sheep with rotational grazing (Molle and Decandia, 2005).

Pasture type	Grass height (cm)	
	Start	**End**
Italian ryegrass	15-20	3-5
Cereals	20-25	8
Lucerne	Start of blooming	3-8
Berseem clover	20-25	8-10
Subterranean clovers and annual medics	10-15	3

Table IT5. Sheep grazing techniques suggested for different grasslands (modified from "Prograze", Molle and Decandia, 2005).

Dry pastures

Prevailing species	Growth phase				
	Stubble (summer)	**Emergence or resprout (autumn)**	**Growth start (winter)**	**Growth end (spring)**	**Beginning of heading/blooming (end of spring)**
Lolium rigidum (self-reseeding)	Moderate grazing to eliminate stubble	Reduce the stocking rate at the emergence of the seedlings to encourage their development up to 3-4 leaves	Continuous or rotational grazing with moderate stocking rate	Continuous or rotational grazing with high stocking rates to delay heading	Avoid intense grazing in order not to reduce re-seeding.
Lolium italicum			Continuous or rotational grazing with moderate stocking rates starting when grass height is 20-30 cm. Avoid grazing if the soil is very humid (compaction)	Continuous or rotational grazing with high stocking rates to delay heading	Continuous or rotational grazing with high stocking rates

Prevailing species	Growth phase				
	Stubble (summer)	Emergence or resprout (autumn)	Growth start (winter)	Growth end (spring)	Beginning of heading/blooming (end of spring)
Annual grasslands (oat, barley, triticale) used for grazing	Grazing "hourly" in the presence of grain to avoid acidosis		Rotational grazing with moderate stocking rates when grass height is 20-30 cm (4-6 weeks post emergence). Residual stubble 8-10 cm. Avoid grazing if the soil is very humid (compaction)	Rotational grazing with high stocking rates to delay heading. Stop when grass height is 5-8 cm	Grazing "hourly" in presence of grain to avoid acidosis
Self-reseeding annual legumes (subterranean clovers, *T. michelianum*, annual medics)	Light grazing to eliminate excess stubble, without depleting the seed bank (min 1.5-2 quintals/ha)	Respect the emergence of seedlings up to 3-5 true leaves	Continuous grazing (preferable) or rotational grazing with moderate stocking rates keeping the height in the range 5-15 cm (avoid shading by grasses)	Continuous (preferable) or rotational grazing with high stocking rates to avoid shading by grasses	Avoid grazing or limit its intensity so as not to compromise seeding, especially with the annual medics and clovers
Annual grasslands (*T. alexandrinum, T. incarnatum*) and sulla grasslands	Sulla: light grazing to eliminate excess stubble	Rotational grazing "hourly" (max. 3 hours) with moderate stocking rates starting from an entrance height of 15-20 cm	"Hourly" rotational grazing with moderate stocking rates when grass height is 20-30 cm. Avoid grazing if the soil is very humid. Leave 8-10 cm of stubble	"Hourly" rotational grazing with high stocking rates. Leave at least 6-8 cm of stubble	"Hourly" rotational grazing with moderate stocking rates
Grasslands based on biennial chicory	Light grazing to eliminate the stubble of adventitious grasses	Respect the beginning of the regrowth up to heights of 15-20 cm	"Hourly" rotational grazing with moderate stocking rates starting when grass height is similar to that of first entry. Avoid grazing if the soil is very humid. Leave 5-8 cm of leaf rosette	"Hourly" rotational grazing with high stocking rates. Leave at least 5-6 cm of leaf rosette	Rotational grazing with moderate stocking rates. Leave 5-6 cm of leaf rosette

▸▸ Grazing systems in the Italian continental regions

In highly productive areas of northern Italy, grazing plays an important role in semi-intensive forage systems under temperate climate (MIPA project, Cavallero *et al.*, 1996). Traditionally in the Po Valley, livestock systems are intensive and rely on annual forage crops that offer high yields per hectare to ensure animal dietary requirements are met and allow a high stocking rate. Nonetheless, in some areas, extensive systems and permanent or temporary grasslands are still the basis of animal feeding, especially where annual forage crops show variable yields, i.e. in sandy, shallow, acidic or silty soils.

Meadows are typical for farms in the area of Parmigiano Reggiano PDO cheese production, as this cheese cannot be produced with milk obtained from dairy cows fed with maize silage. The prevalent conservation systems for meadow forage production are haymaking and haylage. In the case of cow-calf line rearing, the choice of the proper stocking rate is the key to increase farm production. Several experiences showed that a stocking rate of max 3.2-3.4 heads/ha offered the best results in terms of animal weight and herbage quality. An example of a traditional calf-cow line rearing is the Piemontese cow rearing. Piemontese cows show very low milk production and their milking is not convenient. This is why cows, after calving, graze in permanent pastures and their milk is used exclusively by their calves. After weaning, herbage availability in pastures is sufficient to assure to heifers an average daily weight gain of 0.7-0.8 kg/day during a grazing season that lasts for 180 to 220 days with no need for feed integration. The cow-calf line rearing is also advantageous because the number of days between births are reduced (386 days vs. 401 with animals in barns) and the first heat is anticipated (14 vs. 17 months). Other advantages are the reduced workload for farmers (-48%) and a lower use of mechanical means.

The main drawback in these systems to a yearly grazing is the marked seasonality of herbage growth, which is also true in irrigated plains. At the same stocking rate, grassland surfaces needed for grazing increase from 25% to 30% in spring to 100% in late summer.

Some data obtained with temporary grasslands showed the important role of grassland management on the behaviour of forage species. A comparison between binary mixtures based on *Trifolium repens* showed that, in association with *Dactylis glomerata,* the sustainable stocking rate under rotational grazing was higher (+3%) than with *Lolium perenne,* all other factors being equal. However, under intensive continuous grazing, the sustainable stocking rate was higher in mixture with *L. perenne* (+5%). The botanical composition of both mixtures based on *T. repens* varied year after year, because this species tended to increase its relative presence in exclusively grazed and/or in mowed areas. To maintain a balanced mixture, the vigour of the selected cultivar of *T. repens* should be accurately chosen, as well as the nitrogen fertilisation planned in spring.

Other successful mixtures are those based on *Festuca arundinacea*. This species shows a high DM yield, high potential stocking rate, high tolerance to animal trampling and easy haymaking. Unfortunately, the presence of animal dejections on this grass reduced the amount of grazed herbage, especially in the case of continuous grazing (rejected herbage about 30%) compared to rotational grazing (rejected herbage about 20%). The presence of old rejected herbage requires its cut to rejuvenate pastures at the end of each grazing season.

▸▸ Grazing systems in Italian Mediterranean regions

Beef cattle farming system

In extensive breeding, wild resources such as pastures and permanent grasslands are mainly used and complemented with cultivated forage crops (autumn/winter cereal-based temporary grasslands) to fill the gaps that occur during part of the year (Pardini and Rossini, 1996). An example of extensive breeding can be found in the cow-calf line rearing used with the Maremmana cow breed. This breed is rustic and frugal, resistant to diseases and harsh environments and is able to use poor fodder. A stocking rate of 1.3 LSU is used in the most favourable production conditions. Grazing management is intensive continuous grazing. The farming system is planned for the exclusion of animals from grazing on a portion of the farm only during spring season, when forage stocks need to be created (e.g. 1 April to 15-30 June). A part of forage stocks is used as standing hay for deferred grazing in summer. The remaining portion of stocks are mowed at the end of May and the hay is used to fill the late autumn and winter forage gaps. This farming system allows animal dietary requirements to be met year round and requires limited supplies of forage stocks produced on farm from pastures or temporary grasslands. However, in difficult years, when spring forage production is low, the use of extra farm stocks may be needed. In some forage systems, a portion of the natural pasture is replaced by a pure stand meadow of lucerne or a pasture-meadow based on mixtures of perennial grasses (*F. arundinacea* and *D. glomerata*) and legumes (*T. pratense* and *T. repens*).

Dairy sheep farming systems

Two main farming systems are commonly adopted with sheep. The first is the agro-silvo-pastoral system based on wooded grasslands, which are widespread in hilly and mountain areas. The second one is a mixed cereal-animal system, connected to extensive widespread agriculture in the lowland areas and low hilly areas. Nevertheless, a wide range of intermediate situations exists between these two systems, which gives rise to extremely different and complex variations. In practice, each farm is characterised by its own forage system where the grasslands play a different role.

The agro-silvo-pastoral system is mainly based on the use of natural or semi-natural pastures and sowed pastures with or without fertilisation, and a variable proportion of autumn-winter temporary grasslands. The main limitation of this system is the difficulty to match the forage availability from the pastures with animal requirements. In fact, while grass production is concentrated in spring, the highest sheep feed needs are reached in autumn and in winter at the end of pregnancy and the start of lactation. Moreover, the forage productivity of this system is very unpredictable in quantitative and qualitative terms.

Feed integration is used during these two critical periods. At the beginning of autumn, when the first rains occur, the flock is confined to a small plot to allow the re-establishment of pastures, as well as during winter, when the available forage production in pastures is poor. Usually, a relatively high seasonal stocking rate (10 to 20 sheep/ha) is used, especially if there is the need to reserve areas for mowing or standing hay. Sheep are moved to a new plot (pasture or temporary grassland) on the basis of sward production, but more often on the basis of flock milk production: when it starts to decrease, the flock is moved. Pasture grazing covers between 60% and 85% of the total animal requirements.

In the mixed cereal-animal farming system, winter cereals and temporary grasslands, in pure stand or in mixture with annual legumes, represent the main source while the contribution of pasture is restricted to the areas that cannot be mechanised. They are grazed both as green forage and as stubble. In this system type, crop rotation is based on an irregular sequence (from 2 to 4 years) of different cereal crops, such as durum wheat, barley and oat, and fallow pasture. Cereal crops are largely used because they increase the flexibility of this system, which is well suited to the variable Mediterranean climate conditions. In years with adverse meteorological conditions, cereals are only used for grazing and not for haymaking or grain production. In more favourable years, cereals (especially local ecotypes of barley and oat) can provide high-quality DM biomass for grazing during winter (tillering stage) and once grazing is suspended, usually in mid-February, they are used for grain production.

In spring, the flock is confined to a portion of cultivated lands and pastures, where the high grazing pressure is sustained by the rapid spring growth rates of grass. The remaining arable land is used to produce grain and/or hay. One of the main limitations of the mixed cereal-animal farming system is represented by the establishment of cereals, which is strongly conditioned by autumn rainfall as it can cause long delays due to a prolonged summer drought or, on the contrary, waterlogging. The concentration of production, linked to the typical Mediterranean conditions, makes it difficult to identify simplified solutions and requires a wider diversification of forage system resources. The integration of several fodder sources is essential to achieve satisfactory food quality and make the farming system more efficient, flexible and self-sufficient. In this regard, encouraging results have been obtained with the introduction of the annual self-reseeding pasture species and the perennial species for meadow-pasture, which is discussed in the following section.

▸▸ Final remarks on the key aspects for climate change adaptation in Italian Mediterranean grasslands

Several negative effects due to climate change are expected in Mediterranean grasslands: increased failures at establishment, decreased grassland productivity and long-term persistence, shortening of the grazing season unless the grassland is irrigated. Reductions in desirable grassland species is likely in favour of species with low palatability and broad ecological niches, due to reduced competition for water and nutrients (Del Prado *et al.*, 2014). In any case, to prevent these possible negative effects, increasing grassland resilience, improving forage production and rehabilitating permanent grasslands are compulsory. The main factors that can increase resilience and adaptability and which could be considered also as mitigation strategies from climate change are listed below:

- *Sowing annual and perennial species with high summer drought survival.* The predicted changes in rainfall distribution, consisting of relatively lower and more variable autumn rainfall and a shorter spring, mean that some or all of the following traits are needed in annual legumes: (i) earlier maturity for reliable seed set in shorter growing seasons; (ii) more delayed softening of hard seeds to reduce seedling losses from more prevalent false breaks; (iii) greater hardseededness to support grassland survival due to more frequent seasons of little or no seed set; and (iv) a less determinate flowering habit to take advantage of longer growing seasons when they occur. In perennial species, desired characteristics include dormancy or low growth during the drought period, survival across drought periods, and high water use efficiency during the growing season.

- *Increasing legume utilisation.* The biological N-fixing activity of legumes contribute to the soil N-enrichment, and this feature could contribute to farm sustainability. Several species have a different efficiency in fixing atmospheric nitrogen, e.g. up to 150 to 190 kg N/ha/year in sulla and lucerne. In the past, the traditional annual legumes used for grassland rehabilitation were *Trifolium* spp. and *Medicago* spp. Nowadays, other species belonging to the genera *Ornithopus*, *Vicia*, *Melilotus* and *Biserrula* are available on the seed market. Among perennial legumes, lucerne is the most appreciated species in many farming systems but some limitations to its use arise under rainfed conditions, where it shows low forage production, limited persistence and poor tolerance to grazing, requiring the selection of suitable cultivars. Other perennial legumes such as sulla and sainfoin are summer-dormant and already used for both their contribution in stabilising grassland production and forage quality and for their content of condensed tannins, which can promote amino-acid absorption in the intestine as well as reduce the load of gastro-intestinal parasites.

- *Promoting the use of grassland mixtures.* The potential agronomic, environmental and economic advantages of sowing mixtures of forage species and cultivars are widely recognised, especially when mixtures are based on well adapted genotypes. Porqueddu and Maltoni (2007) and Maltoni *et al.* (2007) showed that grass-legume mixtures belonging to different functional groups, achieved higher DM yields, better seasonal forage distribution, better weed control and higher forage quality than pure stands of each species. More persistent grasslands could be also obtained using mixtures of summer-dormant and summer-active perennial species and varieties able to exploit available soil moisture throughout the year (Norton *et al.*, 2012).

- *Benchmarking grassland typologies to improve the management of grassland resources.* Knowledge of grassland typology is needed to adopt the best management practices; in fact, the differences in vegetation and phytosociological associations are still relevant in Mediterranean areas. Agronomic typologies based on the forage value of dominant or reference species, or synthetic indexes were designed in different countries, and recently a first attempt to synthesise and homogenise grassland typologies at plot, farm and regional level in the different EU states was made by Peeters (2015). With regards to grazing, the extent and intensity of grazing differs among vegetation types and geographical locations. Among methods used by technicians and extension services for grassland typology assessment, the pasture-type approach, based on the determination of the pastoral value of grasslands, has been applied in several Mediterranean, Alpine and Apennine areas, with the main goal of characterising pasture vegetation and its potential carrying capacity (Argenti and Lombardi, 2012).

▸▸ References

Argenti G., Lombardi G., 2012. The pasture-type approach for mountain pasture description and management. *Italian Journal of Agronomy,* 7, 39.

Bonifazzi B., Da Ronch F., Susan F., Ziliotto U., 2008. Effects of cutting management on the floristic composition of two permanent Italian ryegrass meadows. *Grassland Science in Europe,* 13, 221-223.

Borreani, Tabacco, 2016. Sistemi foraggeri per l'azienda zootecnica da latte in Pianura Padana: gestione agronomica e costi Available at: http://www.apapd.it/texts/crown2015/Relazione-BORREANI.pdf

Bullitta P., Caredda S., 1982. Influenza degli andamenti climatici sulla reattività del pascolo alla concimazione. *Ann. Fac. Agr. Studi Sa'isarcsi*, XXIX, 131-140.

Caredda S., Porqueddu C., Roggero P.P., Sanna A., Casu S., 1992. Feed resources and feed requirements in the sheep agro-pastoral system of Sardinia. *Proceedings of the IV International Rangeland Congress*, Montpellier, 22-26 April 1991, pp. 734-737.

Cavallero A., Bassignana M., Iuliano G., Reyneri A., 1996. Sistemi foraggeri semi-intensivi e pastorali per l'Italia settentrionale: analisi di risultanze sperimentali e dello stato attuale dell'alpicoltura. In: Attualità e prospettive della foraggicoltura da prato e da pascolo, 221-249.

CLAL. Statistical data. https://www.clal.it/

Del Prado A., Van Den Pol-Van Dasselaar A., Chadwick D., Misselbrook T., Sandars D., Audsley E., Mosquera- EUROSTAT. Statistical data available at: https://ec.europa.eu/eurostat/web/agriculture/data/database

Losada M.R., 2014. Synergies between mitigation and adaptation to climate change in grassland-based farming systems. *Grassland Science in Europe* 19, 61-74.

Huyghe C., De Vliegher A., Van Gils B., Peeters A., 2014. *Grasslands and herbivore production in Europe and effects of common policies*. Editions Quae, Versailles (France).

Lombardi G., Lonati M., Cugno D., 2012. National case studies: Italy. In: Huyghe C., De Vliegher A., van Gils B and Peeters A. (eds.) *Grasslands and herbivore production in Europe and effects of common policies*. Editions Quae, Versailles (France), 178-179.

Maltoni S., Molle G., Porqueddu C., Connolly J., Brophy C., Decandia M., 2007. The potential feeding value of grass-legume mixtures in dry Mediterranean conditions. In: Helgadottir A. and Pötsch E. (eds.). Final Meeting of *COST Action 852*, Raumberg-Gumpenstein (Austria) 30 August – 3 September 2006, 149-152.

Molle G., Decandia M., 2005. Buone pratiche di pascolamento delle greggi di pecore e capre. Available at: http://www.ara.sardegna.it/system/files/documenti/Buone%20pratiche%20di%20pascolamento%20delle%20greggi%20di%20pecore%20e%20capre.pdf

Norton M.R., Lelièvre F., Volaire F., 2012. Summer dormancy in *Phalaris aquatica* L., the influence of season of sowing and summer moisture regime on two contrasting cultivars. *Journal of Agronomy and Crop Science,* 198 (1), 1-13.

Pardini A., Rossini F., 1996. Sistemi pascolivi nell'Italia centro-meridionale. In: *Attualità e prospettive della foraggicoltura da prato e da pascolo*, 251-267.

Peeters A., 2015. Synthesis of systems of European grassland typologies at plot, farm and region levels. *Grassland Science in Europe,* 20, 116-118.

Porqueddu C., Gonzalez F., 2006. Role and potential of annual pasture legumes in Mediterranean farming systems. *Grassland Science in Europe,* 11, 221–231.

Porqueddu C., Maltoni S., 2007. Biomass production and unsown species control in rainfed grass-legume mixtures in a Mediterranean environment. In: Helgadottir A. and Pötsch E. (eds). *Proceedings of the COST 852 final meeting,* 30 August-3 September 2006, Raumberg-Gumpenstein, Austria, 41-44.

Porqueddu C., Ates S., Louhaichi M., Kyriazopoulos A.P., Moreno G., del Pozo A., Ovalle C., Ewing M.A., Nichols P.G.H., 2016. Grasslands in 'Old World' and 'New World' Mediterranean-climate zones: Past trends, current status and future research priorities. *Grass and Forage Science,* 71, 1-35.

Porqueddu C., Melis R.A.M., Franca A., Sanna F., Hadjigeorgiou I., Casasús I., 2017. The role of grasslands in the less favoured areas of Mediterranean Europe. *Grassland Science in Europe*, 22, 3-21.

Volaire F., Barkaoui K., Norton M., 2013. Designing resilient and sustainable grassland for a drier future: adaptive strategies, functional traits and biotic interactions. *European Journal of Agronomy* 52, 81-89.

Zarovali M.P., Yiakoulaki M.D., Papanastasis V.P., 2007. Effects of shrub encroachment on herbage production and nutritive value in semi-arid Mediterranean grasslands. *Grass and Forage Science*, 62 (3), 355-363.

Imprimé pour vous par Books on Demand (Allemagne)